Cultures and Contexts:
the Caribbean

SECOND EDITION

Aisha Khan

NEW YORK UNIVERSITY

LINUS
Learning

Published by Linus Learning

Ronkonkoma, NY 11779

ISBN 10: 1-60797-576-9

ISBN 13: 978-1-60797-576-2

Printed in the United States of America.

This book is printed on acid-free paper.

Print Number   5   4   3   2   1

Table of Contents

iv |

Acknowledgments

Syncretism in religion a reader by LEOPOLD, ANITA Reproduced with permission of TAYLOR & FRANCIS GROUP LLC-BOOKS in the format Photocopy for a coursepack via Copyright Clearance Center.

Global transformations: anthropology and the modern world by TROUILLOT, MICHEL-ROLPH Reproduced with permission of PALGRAVE MACMILLAN in the format Photocopy for a coursepack via Copyright Clearance Center.

The birth of Caribbean civilisation: a century of ideas about culture and identity, nation and society by Bolland, O. Nigel Reproduced with permission of JAMES CURREY LTD. in the format Photocopy for a coursepack via Copyright Clearance Center.

Anthropological quarterly by CATHOLIC ANTHROPOLOGICAL CONFERENCE; CATHOLIC UNIVERSITY OF AMERICA Reproduced with permission of INSTITUTE FOR ETHNOGRAPHIC RESEARCH

Sun, sex, and gold: tourism and sex work in the Caribbean by KEMPADOO, KAMALA Reproduced with permission of ROWMAN & LITTLEFIELD PUBLISHING GROUP, INC. in the format Photocopy for a coursepack via Copyright Clearance Center.

Beyond sun and sand: Caribbean environmentalisms by Baver, Sherrie L.; Lynch, Barbara D. Reproduced with permission of RUTGERS UNIVERSITY PRESS in the format Photocopy for a coursepack via Copyright Clearance Center.

Race, nature, and the politics of difference by Moore, Donald S.; Kosek, Jake; Pandian, Anand Reproduced with permission of DUKE UNIVERSITY PRESS in the format Photocopy for a coursepack via Copyright Clearance Center.

The Journal of Pan African studies by California Institute of Pan African Studies Reproduced with permission of JOURNAL OF PAN AFRICAN STUDIES/ JPAS

Indentured labor, Caribbean sugar: Chinese and Indian migrants to the British West Indies, 1838-1918 by LAI, WALTON L. Reproduced with permission of JOHNS HOPKINS UNIVERSITY PRESS-BOOKS in the format Photocopy for a coursepack via Copyright Clearance Center.

The coolie speaks: Chinese indentured laborers and African slaves in Cuba by YUN, LISA Reproduced with permission of TEMPLE UNIVERSITY PRESS in the format Photocopy for a coursepack via Copyright Clearance Center.

Maharani's misery: narratives of a passage from India to the Caribbean by Shepherd, Verene Reproduced with permission of JAMCOPY, JAMAICAN COPYRIGHT LICENSING AGENCY in the format Photocopy for a coursepack via Copyright Clearance Center.

CHAPTER ONE

Africa, Europe, and Asia in the Making of the 20th Century Caribbean

AISHA KHAN

The Caribbean of today began to form half a millennium ago, impelled by European colonial expansion harnessed to nascent capitalism and centered on resource extraction and sugar plantations producing for a global market. Within 50 years of Columbus's landing, indigenous Caribbean populations had been dramatically reduced, largely due to disease and the harsh conditions of labor imposed by the Spanish colonizers. This diminution of indigenous peoples was accompanied by the addition of foreigners from the "Old World" of Europe, Africa, and later Asia—a socially engineered assemblage of disparate ethnolinguistic groups under conditions of coerced labor and massive wealth accumulation. The imported groups included indentured Europeans, enslaved Africans, and, later, indentured Africans and Asians.

The transformations of the plantation system had various effects on the racial and demographic composition of different colonial territories. For example, the Hispanophone Caribbean, particularly Cuba and Puerto Rico, was not significantly developed for the global sugar market until the 19th century (although by mid-century Cuba and Puerto Rico had emerged as the first and third largest producers of sugar in the hemisphere), and the proportion of

European populations compared to non-European populations was far greater there than in the Francophone and Anglophone colonies.

Over the 19th century, slavery was gradually abolished in the Caribbean. Newly independent Haiti (formerly Saint-Domingue) abolished slavery in 1804, followed by the British West Indies in 1838, the French possessions in 1848, all Dutch territories by 1863, and Cuba in 1886. Emancipation presented plantation owners with a dilemma: ensuring sugar and other production at high levels without the benefit of enslaved labor, or with diminishing numbers of freed workers willing to engage in plantation labor under the conditions offered by the plantocracy. One strategy implemented by Britain and France was that of freeing Africans from the slave trade of other European colonizers (Dutch, Spanish, Portuguese) and then sending them to British and French Caribbean colonies as indentured laborers. Almost 40,000 Africans were thus sent to the British West Indies and approximately 16,000 to the French West Indies (Schuler 1980).

Another form of 19th-century indenture brought immigrant laborers from Asia into the region. Organized as either state projects or private enterprises, indenture schemes evolved over eight decades and changed the demographic, cultural, and social terrain of the Caribbean as irrevocably as African slavery had done earlier. Between 1890 and 1939, for example, the Dutch recruited almost 33,000 Javanese, primarily from Central Java and Batavia, for their Caribbean colony of Suri-name. The two principal source regions of indentured labor, however, were India and China. Itself a British colony, India experienced indenture as a government-regulated industry, with laborers recruited primarily from the regions of Oudh, Bihar, and Uttar Pradesh and shipped out from the ports of Calcutta and Madras. Between 1838 and 1917, almost 400,000 Indians arrived in the British Caribbean, the majority in Guyana and Trinidad. Although China was never colonized, its political vulnerability allowed private interests to orchestrate indenture schemes, largely from Canton. Between 1840 and 1875, approximately 142,000 indentured Chinese arrived in Cuba (Helly 1993, 20); from 1853 until 1866 and in trickles thereafter, about 18,000 Chinese were indentured in the British West Indies (Look Lai 1993, 18). Later—beginning around 1890, and concentrated between 1910 and 1940—a second wave of Chinese immigrants, this time not under indenture, arrived in the Caribbean.

Figure 27.1 Newly arrived Indian laborers in Trinidad. Photograph (1897).

The relationships of Asian indentured laborers with the local populations they encountered have influenced the values, identities, and cultural practices of their respective societies. To one extent or another, all the Asian immigrants were initially viewed by the locals as labor competition. Particularly where they constitute a large percentage of the population, Indians have been represented by local anti-indenture interests as "scab" labor, yet historically they also have been pitted against Afro-Caribbean workers. The tensions arising from perceived and actual labor conflicts have left a monumental legacy of racial politics in such contemporary societies as Guyana and Trinidad, where Indians represent more than 40% of the population. Perhaps because of their relatively smaller numbers, Chinese and Javanese laborers have had less fraught relationships with established populations, especially with those in similar occupational and class positions. In Cuba, for example, Chinese indentured laborers worked side by side with enslaved Africans. Enmity between these two groups was encouraged by colonial authorities as a divide-and-rule strategy, but tensions expressed in racial terms did not significantly persist into the present, either in Cuba or in other parts of the region. Once the Chinese found their economic niche primarily in the retail trades and shopkeeping, they no longer represented labor competition to other populations.

Migrants to the Caribbean from the Levant—known as "Syrians," "Syrian-Lebanese," or *árabes*—also began to arrive in the

1860s, increasing their numbers significantly by the 1890s. Most were Maronite Christians leaving Ottoman-occupied regions. Lebanese immigrants came first, followed by Syrians and Palestinians. Although they spread out across the Caribbean (and into Latin America, where they are also called *turcos)*, certain communities predominated in particular countries. For example, of the three groups from the Levant, Lebanese comprise the largest population in Jamaica and the Dominican Republic, and Palestinians in Haiti (Nicholls 1980). These immigrants came as individuals, or sometimes in families, rather than in an organized migration arrangement; over the years, other family members followed. Although a few went into agricultural production, others became itinerant peddlers. Within a few generations these communities branched out into import-export trading, and today they comprise a large population of affluent and politically active citizens.

In addition to the transcontinental migrations, intraregional population movements have been crucial in contributing to the character of today's Caribbean. Although interisland labor migrations commenced soon after emancipation in the British West Indies, the late 19th century and first decades of the 20th saw the most dramatic population movements within the Caribbean basin. For example, between 1900 and 1914, some 60,000 Barbadians labored on the Panama Canal. Likewise, between 1917 and 1931, some 300,000 Jamaicans, Haitians, and other labor migrants from the region worked in Cuba on sugar plantations and in factories (De la Fuente 2001, 102), and several thousand more left the Leeward Islands to work in the US-owned sugar industry in the Dominican Republic (Conway 2003, 339). From the 1880s, Haitians crossed into the Dominican Republic to work on foreign-owned plantations as well as to farm smallholdings along the border. By 1935, Haitians in the Dominican Republic numbered perhaps 200,000—more than 10% of the national population (Andrews 2004, 140). Aside from the economic dimension of these population displacements, the resulting cultural and linguistic exchanges contributed significantly to the continuous formation of new social fabrics.

INTERPRETING "CREOLE" SOCIETIES

Four major languages are spoken in the Caribbean: Spanish, English, French, and Dutch. The 17 Caribbean countries that are predominantly Anglophone comprise more than 17% of the region's population, yet the total English-speaking population of the Caribbean is less than that of the Dominican Republic alone. These statistics clarify the demographic predominance of the Spanish-speaking countries of Cuba, Puerto Rico, and the Dominican Republic, which represent 61% of the Caribbean population. Of the 20% of Caribbean peoples who speak French or variations of French, three-quarters live in Haiti. The Dutch speakers of Suriname and the Netherlands Antilles represent another 2% (Knight 1995, 34). Other languages, spoken by fewer numbers of people, include Hindi and Javanese.

The languages of the European colonizers remain the official languages of formal Caribbean education and legal systems, but numerous African languages brought by the slaves fused with European, Asian, and Amerinidian languages to create numerous "creole" languages, which are the spoken vernaculars of everyday life in a number of Caribbean countries. Most Caribbean creole languages are young as languages go, having existed for not more than two or three centuries. Today, however, there are growing written literatures in creole languages, and movements to promote the languages to equal standing as vehicles of formal instruction and communication. Among the most familiar examples is Haitian Kreyol, the spoken language of approximately 12 million insular and diasporic Haitians, which along with French has been an official language in Haiti since 1961. Other widely spoken Creoles include Jamaican patois, which is spoken by about four million people in and outside Jamaica, and the patois of Trinidad and Tobago, a historical legacy primarily of French on Trinidadian English, which has been in decline since about the mid-20th century. In Suriname, Sranan Tongo is the language of approximately 300,000 people; in Aruba, Bonaire, and Curacao, Papiamento is spoken by more than 350,000. And although the varieties of Spanish spoken in Cuba, Puerto Rico, and the Dominican Republic share a number of linguistic properties, they also have discernable differences based on geographic location and local histories.

As a region whose very foundations lie in multiple origins — assorted languages, varied religions, diverse worldviews, contrasting cultural traditions — the Caribbean has long been represented by observers, local and international alike, as the epitome of heterogeneity (Trouillot 1992; Glissant 1995). For more than a century, the region has been the object of attempts to explain cultural change over time as phenotypically and culturally heterogeneous peoples come into what is commonly known as "culture contact" and undergo the cultural transformations that such contact engenders. The questions about how cultures retain or lose continuity across vast geographical spaces (for example, after transatlantic migrations) and over extended periods of time (from the colonial period to the present), how the dynamics of unequal power relations foster or challenge cultural assimilation in new environments, and how identities and worldviews are forged in the process remain central to the study of Caribbean cultures. And certain concepts that have emerged from this study — creole, creolization, *creolite*, survivals, retentions, transculturation, and syncretism — have achieved much broader usage.

From its colonization, the Caribbean has represented newness, which Europeans captured in the term "creole." When applied to the region, the Spanish word *criollo* and the Portuguese word *crioulo* (derived from the verb *criar*, "to raise or bring up") signified something or someone originating in Europe (or Africa) and reproducing itself in the New World. Thus animals, plants, and people could all be designated as creole. Creole people were the descendants of Europeans or Africans born in the Caribbean, as well as the offspring of African and European parents. Inherent in the idea of creole identity was an assumption that being born in the Caribbean or being the "mixed" descendant of two racially differentiated parents meant losing one's ancestral cultural heritage.

Many of the earliest and most important social science studies of Creole identity in the Caribbean were concerned specifically with Afro-Caribbean populations. They framed the question of cultural change over time in terms of the search for cultural heritage through "survivals" (or "retentions") and reinterpretations. Heralded in North America by anthropologists Melville and Frances Herskovits, this approach emphasized empirical evidence: the prevalence and intensity of survivals — as perceived by the Herskovitses in, for example, art forms, cuisine, technology, language, and religion — could prove cultural continuity between Africa and the Americas. As debates about African survivals progressed, however, the emphasis shifted from observable traits in the identification and study of cultural forms toward values, style, and systems of relationships.

While retaining an assumption that Creole identity involves the loss of former cultural heritage in the process of forging new cultural adaptations, this shift brought greater interest in the variations of cultural forms and identities produced from dissimilar types. Thus the question of how new, Creole types were produced became as much a focus of study as the types themselves. The creation of models to explain these processes emerged from different thinkers throughout the His-panophone, Anglophone, and Francophone Caribbean. Among the most influential of these were Fernando Ortiz's early 20th-century model of *transculturación* based on Cuba, Edward Kamau Brathwaite's mid-20th-century model of creolization derived from Jamaica, and Jean Bernabe, Patrick Chamoiseau, and Raphael Confiant's (1993) late 20th-century model of *creolité* based on Martinique.

Ortiz coined the term "transculturation" as a way to interpret the multiplicity of histories, cultures, languages, religions, and worldviews that collectively form the Caribbean, and to express the various phases of the processes of transition from one culture to another. Ortiz saw these processes as entailing more than the simple and passive acquisition of, or submission to, other cultures, which he equated with "acculturation." Instead, transculturation involved the simultaneous loss or displacement of a preceding culture ("deculturation") as well as the resultant creation of new cultural phenomena ("neoculturation") (Ortiz 1995, 102-3). Parsing the concept into processes, or active elements, allows a highlighting of the ways in which subjugated peoples create their own versions of the dominant culture.

In the Anglophone Caribbean, Brathwaite's analysis of what he called the "creole society" of Jamaica emphasized the creation of new forms through the synthesis of existing ones. Arguing against understanding black and white populations as "separate nuclear units," Brathwaite saw them as being "contributory parts of a whole" that produce a uniquely Caribbean culture. Creolization here represents the potential for social integration and unity, where the "mixed" population serves "as a bridge, a kind of social cement" that integrates society (Brathwaite 1971, 307, 305). In calling for a renewed emphasis on creole identity and the literary value of the creole language, the most recent Francophone *creoliste* writers and activists celebrate the heterogeneous dimensions that together comprise the Caribbean or, in the words of Martinican poet and writer Edouard Glissant, constitute *Antillanité* (Caribbeanness). The *creoliste* position, along with those of other thinkers, points to the abiding debates about how to characterize and give meaning to the forms of diversity so apparent in the region.

IDEOLOGIES OF RACE, COLOR, AND CLASS

From the earliest days of colonial rule, the Caribbean social and moral order was based on ranked gradations of "races" and "colors" represented by such physical attributes as skin color, hair texture, and facial features. These criteria were treated as literal descriptions of appearance, and their presumed fixed qualities formed a hierarchy of identities—from "white" at the top to "black" at the bottom, with various mixtures and gradations in between—supported by legal structures as well as social values and mores. Consequently, for much of Caribbean history, race and color also have connoted social position and class status. Yet the recognition of a vertical color continuum separates the Caribbean from the rigid binary racial logic of the United States.

Given the legacies of colonial rule and ideology, color and race are still commonly used in daily conversation as idioms for social organization. In Jamaica, for example, the color term "brown" (or "colored") serves as a category of racial identity but also connotes middle-class status. Color terms are necessarily relational; being "white" or "brown" or "black" necessarily means not being something else. In Haiti, *mulâtre* is an in-between term connoting a mixture of "black" and "white," flexible in its interpretation yet typically positioned above "black" and below "white." In the Dominican Republic, *indio* literally translates as "Indian," suggesting indigenous heritage, but its contemporary application signifies a lighter skin color (and perhaps straight hair)—someone not "black," yet also not "white." In Martinique, *beke* refers to French "white" slave owners and their descendants. "Trinidad white" and "French creole" have served as categories of racial identity in Trinidad, specifically distinguished from British, French, and Spanish "whites," who, in this racial accounting system, historically could claim to be "pure" white and, concomitantly, members of the upper classes. In Trinidad, the term "red" generally refers to a light-skinned individual of mixed "black" and "white" parentage (positioned toward the upper-status end), while in Barbados it is also a historical reference to "red legs" communities—poor whites who, from the days of the slave plantation, labored outdoors and hence were likely to get sunburned.

Mixedness can also refer to multiple combinations, not simply the amalgamation of "black" and "white." Thus, in the Francophone Caribbean, the term *marabou* refers to a black-white-Amerindian combination. In Trinidad the term "Spanish" should be interpreted as if in quotation marks, indicating a particular and fluctuating combination of local criteria, including area of origin (Venezuela, or certain locations in Trinidad with historical concentrations of Spaniards, Amerindians, and Venezuelan immigrant labor), skin color (some variation of "brown" or "red"), hair texture (not curly), and self-ascription (Khan 1993).

Notably, these terminologies are based on an African-European axis: the hierarchical color continuum does not lexically include South Asians or Chinese, or the mixed offspring of South Asian or Chinese and European parents. Though the term *achinado* is used in Cuba to index Chinese phenotypical features (as, for example, in *mulato achinado)*, there is only one term, *dougla*—common in Guyana and Trinidad—indicating individuals of mixed South Asian and African descent. *Indio* (Amerindian) in the Hispanophone Caribbean and "Spanish" or "French Creole" in Trinidad are not color terms per se, but are measured along the continuum of black and white ancestry. "Indian" (South Asian), "Chinese," and "Syrian-Lebanese" in the Anglophone Caribbean, "Hindustani" in the Dutch Caribbean, and *Hindou* in the Francophone Caribbean are common categories not amalgamated into the black-white lexicon.

Twentieth-century anticolonial movements encouraged Caribbean societies to project themselves as modern, sovereign democracies. Race and color were thus applied by political leaders to nationalist projects, with local perceptions of cultural, racial, and color heterogeneity representing the ideal of "unity in diversity": multicultural tolerance and harmony. In this mid-century nationalist narrative, evident throughout the region, evoking creolization became tantamount to celebrating the strengths of cultural heterogeneity. The claimed character of such "creole" societies as Trinidad, Guyana, Suriname, Jamaica, Belize, Cuba, Martinique, Curacao, and French Guiana is a national identity in which culturally and racially distinct groups cooperatively coexist as united, independent nations. Claims about the strengths of diversity are deliberated across the region, and the discussion largely reflects debate about representation—that is, which constituent cultural-racial groups will be at the forefront of defining national identity and, by implication, national interests. The lack of any uniform local understanding of what constitutes "diversity" across the region complicates these claims and the narrative of harmony; each nation-state's self-perception is configured somewhat differently in terms of its cultural, religious, racial, and ethnic composition. Thus nationalist projects tackle the broad Caribbean theme of multiplicity in historically and ideologically particularistic ways. Whatever direction these discussions take, however, they show no indication of muting the salience of racial accounting and color categories in the region.

GENDER AND KINSHIP

Caribbean kinship forms have been shaped by the different governing structures and sociocultural character of European colonizers. In the British-influenced Commonwealth West Indies, for example, social welfare policy played a significant role in shaping family life, particularly during the period of high social unrest between the two world wars. Throughout the wider Caribbean, however, kinship is understood to have certain common patterns that transcend specific colonial identities. These patterns derive largely from two contexts that share certain region-wide similarities: slave plantation society, and class stratification systems that arose after emancipation. Because all the populations associated with slavery are African in origin, for example, kinship practices among Afro-Caribbeans have generally been distinguished in terms of social class positions rather than in terms of the ethnic, cultural, or racial differences that exist among these populations. Such differences come into play only when kinship practices are compared across the ethnoracially, culturally heterogeneous groups of the region, usually based on divisions among "African," "European," and "Asian" (or "East Indian") peoples.

In slave plantation societies, West African kinship forms met the constraints and agendas of European colonial rule. Although each colony had legal structures to control slaves' cultural practices, patterns of conjugality and religious worship largely remained outside the purview of the master's control. Over time, many forms of mating and cohabitation—from legal marriage to extra-residential unions, with a range of arrangements in between—became part of the social landscape. Such flexible diversity in domestic forms contributed significantly to the idea that the Caribbean consisted of, at best, fluidly adaptive social institutions and, at worst, unstable ones—in alleged contrast to European and North American society. Until the late 20th century many policy makers, scholars, and other observers assumed that among Afro-Caribbean peoples, corporate kinship groups were less important than dyadic ties between individuals. Today these assumptions are under more careful scrutiny, and the gender dynamics of Caribbean kinship are primarily understood as constructive responses to such shifting forces as migration, educational opportunity, and labor conditions.

In the context of post-emancipation society, the values and behaviors associated with different classes reflect a status hierarchy. According to a model first elaborated by anthropologist Peter Wilson (1973), Afro-Caribbean women are judged on their "respectability," which reflects bourgeois European colonial values, while Afro-Caribbean working-class men are judged on their "reputation," which reflects local or creole values and a disinclination to emulate European values and standards of behavior (nuclear families, wage employment, formal education, obedience to social and legal norms, self-control). Reputation is meant to convey an alternative system of status and prestige, one that emphasizes verbal dexterity, ease and mobility in the public sphere, sexual prowess, and lack of restraint in engaging in these behaviors. Although this model has been critiqued for its oversimplified binary opposition of gender roles—for example, Caribbean women engage in their own practices of "reputation"—it remains useful for understanding the ways in which gender in the Caribbean is culturally expressed and ideologically linked to class divisions.

Along with class differences, ethnocultural distinctions are significant in the formation and representation of Caribbean kinship practices. Like Afro-Caribbean peoples, Indo-Caribbeans who settled in the region were a diverse population. They came from various parts of the Indian subcontinent, spoke several languages (notably Hindi, Urdu, and a dialect called Bhojpuri), belonged to different religions (primarily Hinduism and Islam), and had a number of family and household arrangements. But Indo-Caribbean populations differed from their Afro-Caribbean and other neighbors in some key respects. Whereas plantocracies throughout the region discouraged enslaved Africans from practicing their cultural traditions, colonial authorities in the age of the indentured labor system were more flexible regarding Indian immigrants' activities. Practices and observances that were thought to hinder plantation production were curtailed, but cultural life was far freer than under slavery. Yet Hindu and Muslim marriages were not legalized in the Caribbean until the mid-20th century, presenting challenges for Indo-Caribbeans not faced by their neighbors who practiced Christianity or Afro-European religions. Moreover, marriages between Indo-Caribbeans reflected cultural traditions brought from the subcontinent—that is, they were often arranged, with families seeking potential mates for their children through the services of matchmakers from the community. This practice continues today, though less formally and much less frequently.

COSMOLOGIES AND BELIEF SYSTEMS

The basic distinction made between religions local to the Caribbean and those derived from outside the region raises an interesting question about the meaning of "indigenous." Because the first peoples to inhabit the region, Amerindians, succumbed very early to European conquest, there are no indigenous religions in the Caribbean, in the sense that the term is used to describe religions of native peoples in Latin and North America. Rather, all Caribbean religions have undergone transformation over time and derive from predecessor religions that were variegated in their belief and practice. Yet because of the legacy of cultural creolization, the Caribbean represents a major crucible of Creole, or syncretic, religions.

As in other parts of the world, religion has offered the peoples of the Caribbean a way to interpret and engage past and present social conditions and forms of inequality. Among the most well known examples is the role of vodou priests and priestesses in slave revolts, such as Boukman's 1791 insurgency, which is thought to have initiated the Haitian Revolution. Similarly, in 1884, in response to British colonial curtailments, the insistence of Trinidadian Muslims and Hindus on carrying on Hosay (the Caribbean version of Muharram, the Shi'a Muslim ritual mourning the martyrdom of the Prophet Mohammed's grandson, Imam Hussain) resulted in Trinidad's "Hosay Riots."

At the same time, Caribbean religions offer alleviation of natural and supernatural distress, notably problems of health, success, and fidelity. For example, brujería in Puerto Rico—a blend of popular Catholicism, Afro-Latin religions, French spiritism, and folk Protestantism—engages in healing, advocacy, and solving both metaphysical and practical problems among populations who have few alternatives or who avail themselves of a number of religious options (Romberg 2003). Although they possess their own distinctive histories, characteristics, and modes of practice, vernacular religions such as Haitian vodou, Cuban Santería, and Trinidadian ori-sha also serve such needs. And institutionalized forms of religious practice such as Christianity, Islam, Judaism, and Hinduism have found fertile ground in the region. Associated with European colonizers, the Catholic and Protestant churches have worked to conserve their formal traditions and doctrines even under the forces of transformation and syncretism. Hindu and Muslim communities in the region have sponsored missionaries and educators from India since the mid-19th century and from Pakistan in the 20th.

Caribbean religions are among the most complex examples of the emergence and transformation of cultural lifeworlds in the Americas. Given their numerous sources and formations, and their tendency to eschew orthodox axioms in favor of heterodox practices guided by a few broad principles, religions emerging from the Caribbean are characterized by amalgamation and recombination. Added to syncretic or creole religions deriving from the Caribbean context are religions whose doctrines and belief systems, themselves varied and changing over time, derive from "Old World" origins. Thus, today even a cursory list of religions in the region would be long— Catholicism, Protestantism, evangelical and Pentecostal movements, Judaism, Hinduism, vodou, Santería, Islam, espiritismo, Rastafari, and ori-sha—made even longer by a number of demographically smaller but socially significant traditions such as Kali worship in Guyana, brujería and Mita worship in Puerto Rico, Quimbois in Martinique, and Winti in Suriname.

Equally important are historical and contemporary magical practices (often subsumed under the term "obeah") that involve supernatural powers, deriving largely from West African divination and healing practices and, to a lesser extent, Hindu and Christian cosmologies. The meaning of obeah has changed over the centuries. Among 17th- to 19th-century Africans and Afro-Caribbeans it was associated with salutary objectives, such as alleviating illness, protecting against harm, and avenging wrongs. Euro-colonial and local bourgeois ideologies emphasized the dangerous aspects of obeah, often equating it with Judeo-Christian interpretations of evil forces. Often, positive and negative assessments existed simultaneously, making local opinion about obeah ambiguous. Today, as in earlier eras, its practice represents tensions between the ways in which practitioners interpret obeah's methods and objectives, and the ways in which those methods and objectives are perceived by outsiders.

Caribbean religions are expressions of traditions of creativity, resistance, and flexibility that continuously build on as well as disassemble older and current forms of knowledge, heritage, and custom. The challenge in understanding them is to grasp that difference and similarity exist at the same time. Hinduism, as practiced by the progeny of indentured laborers, reflects both the remembered traditions that early immigrants brought with them from India and a contemporary global Hinduism that travels across the Hindu diaspora. While Caribbean Hindus may interpret their forms of worship as replicating those in India, they also recognize that certain transformations and syncretisms have occurred for almost 170 years in the Caribbean.

In contrast, Rastafari's origins are in Jamaica, where religious movements based in Afro-Caribbean folk Christianity, the pan-Africanism of Marcus Garvey, grassroots reinterpretations of the Old Testament, and the veneration of Haile Selassie of Ethiopia coalesced in the 1930s, giving rise to the religious, philosophical, and political worldview of today's Rastafari movement. In it, Africa plays a great symbolic role as a place of desired return and the antithesis of "Babylon"—all places and forms of consciousness in which predatory relationships and "mental slavery" abound. Yet although thus memorialized, Africa is not literally remembered by many Rastafari, the vast majority of whom have never had direct experience with societies and cultures in Africa or Ethiopia (two terms often used synonymously). Nonetheless, Africa/Ethiopia represents for them an indispensable emblem of unity, self-determination, authenticity, and morality.

Islam, meanwhile, first came to the Caribbean as the religion of some African slaves. With the advent of indentured laborers from India, Islam gained an increased presence in the region. Notable today are the numerous *masjids* (mosques) that dot the landscape of many countries, from Trinidad to Guyana, Puerto Rico, and Suriname. Some masjids are humble, built to serve small communities and local villages; others are grand, built as centers of learning as well as centers of worship for larger populations in the towns and cities. In these places of worship that serve *jamaats* (congregations) large and small, imams (religious leaders) work to preserve the *Sunnah* (Muslim way of life). At the same time, Islam in the Caribbean encapsulates the simultaneous inclusiveness and exclusions of a religion claimed by different ethnic groups, practiced according to divergent interpretations of doctrine, and, in certain contexts, participated in by non-Muslims. This is perhaps best seen in the ritual of Hosay, the Caribbean version of Shi'a Islam's commemoration, Muharram.

Figure 27.2 A tadja at Hosay in St. James, Trinidad. Photograph by Dr. Ted Hill (1950s).

Historically spread throughout the Anglophone Caribbean, today Hosay is practiced on a major scale only in Trinidad, where it is simultaneously an important religious event, a freighted political statement, an embattled heritage claim, and a multicultural symbol. Mourners of Hussain march with enormous, elaborate representations of the *tadjas* (*tazzias,* or representations of the martyrs' tombs; see fig. 27.2). This procession has been treated by some local participants less like a sacred commemoration than like a parade, where music and general revelry may occur on the sidelines. Despite its Muslim origins, Hosay in Trinidad also has always involved Hindus and Afro-Trinidadians. Hindus have long been key participants in the building of the tadjas, and Afro-Trinidadians traditionally have played a significant role as drummers as well as bearers (along with Hindu and Muslim Indo-Trinidadians) of the tadjas in procession. Moreover, Hindus sometimes make their own vows and offerings during Hosay. This ritual was the only significant element in the Indian cultural repertoire that provided a social bridge to the rest of 19th-century Trinidadian society (Singh 1988, 4). Given its multiple interpretations and diverse participants, Hosay lends a distinctive religious and cultural tenor to Trinidad's national culture. The combination of participants and their varied forms of involvement has given rise to debates among Muslims and non-Muslims about the authenticity of Hosay and its appropriateness in Islam. Other observers argue that this ceremony's heterogeneity and cooperation counters the divide-and-rule antagonism among subordinate groups (notably Afro- and Indo-Caribbeans) encouraged by British colonizers, offering a natural space for a Creole unity.

Religion is just one of innumerable examples of the ways in which Africa, Europe, and Asia have together produced the 20th-century Caribbean. In the organization of labor, language, group identities, and kinship as well as religion, these Old World continents inspired the creation of many multidimensional New World cultures and societies. The productive relationship between "old" (existing) and "new" (emerging) that gave rise to the Caribbean of today must be understood as a consequence of the protracted and often painful tension between domination (initiated by the articulation of colonialism and capitalism, which significantly defined the region) and resistance (local forms of accommodation and challenge) to that domination. The 20th-century Caribbean represents one of the most diverse places on earth; this diversity is richly symbolic of the workings of the human imagination in both felicitous and forbidding circumstances.

WORKS CITED

Andrews, George Reid. 2004. *Afro-Latin America, 1800-2000.* New York: Oxford University Press.

Bernabe, Jean, Patrick Chamoiseau, and Raphael Confiant. 1989. *Eloge de la creolite.* Paris: Gallimard.

Brathwaite, Edward Kamau. 1971. *The Development of Creole Society in Jamaica, 1770-1820.* Oxford: Clarendon Press.

Conway, Dennis. 2003. "The Caribbean Diaspora." In *Understanding the Contemporary Caribbean,* edited by Richard S. Hillman and Thomas J. D'Agostino, 333-53. Boulder, CO: Lynne Rienner.

De la Fuente, Alejandro. 2001. *A Nation for All: Race, Inequality, and Politics in Twentieth-Century Cuba.* Chapel Hill: University of North Carolina Press.

Glissant, Edouard. 1995. "Creolization in the Making of the Americas." In *Race, Discourse, and the Origin of the Americas,* edited by V. L. Hyatt and R. Nettleford, 268-75. Washington, DC: Smithsonian Institution Press.

Helly, Dorothy, ed. 1993. *The Cuba Commission Report: A Hidden History of the Chinese in Cuba.* Baltimore: Johns Hopkins University Press.

Khan, Aisha. 1993. "What is 'a Spanish'? Ambiguity and Mixed Ethnicity in Trinidad." In *Trinidad Ethnicity,* edited by Kevin Yelvington, 180-207. Knoxville: University of Tennessee Press.

Knight, Franklin. 1995. *Race, Ethnicity, and Class: Forging the Plural Society in Latin America and the Caribbean.* Waco, TX: Markham Press Fund.

Look Lai, Walton. 1993. *Indentured Labor, Caribbean Sugar.* Baltimore: Johns Hopkins University Press.

Nicholls, David. 1980. *Arabs of the Greater Antilles.* New York: Research Institute for the Study of Man.

Ortiz, Fernando. 1995. *Cuban Counterpoint: Tobacco and Sugar.* Durham, NC: Duke University Press.

Romberg, Raquel. 2003. *Witchcraft and Welfare: Spiritual Capital and the Business of Magic in Modern Puerto Rico.* Austin: University of Texas Press.

Schuler, Monica. 1980. *"Alas, Alas, Kongo": A Social History of Indentured African Immigration into Jamaica, 1841-1865.* Baltimore: Johns Hopkins University Press.

Singh, Kelvin. 1988. *Bloodstained Tombs: The Muharram Massacre 1884.* London: Macmillan Caribbean.

Trouillot, Michel-Rolph. 1992. "The Caribbean Region: An Open Frontier in Anthropological Theory." *Annual Review of Anthropology* 21:19-42.

Wilson, Peter. 1973. *Crab Antics: A Caribbean Case Study of the Conflict between Reputation and Respectability.* New Haven: Yale University Press.

CHAPTER TWO

The Caribbean: Marvelous Cradle-Hammock and Painful Cornucopia

CARLOS GUILLERMO-WILSON

Translated by Elba D. Birmingham-Pokorny and Luis A. Jimènez

In 1492, the three caravels—the *Santa Maria*, the *Pinta*, and the *Niña*— landed on the coast of the island of Quisqueya, where later Santo Domingo, the oldest Spanish colony in the New World, was founded. This event was the beginning of an impressive historical Caribbean phenomenon: a marvelous cradle-hammock and painful cornucopia.

The ceremonies of the quincentennial (1492-1992) of that historic October 12, jubilantly celebrated the marvel born in that cradle, or better yet, Caribbean "cradle-hammock." In the fifteenth century, the legends of El Dorado and the Fountain of Youth called attention to the marvelous and obsessive search which started in the Caribbean, and consequently caused much inrerest in Spain in the news about Tenochtitlán, the great center of the Aztecs; Chichen Itza, a great center of the Mayas; Darién and die Pacific Ocean (then known as the South Sea); and Cuzco, the great center of the Incas. One of the most important legacies of the marvelous cradle-hammock is the Spanish spoken in the Caribbean.

In 1492 Antonio de Nebrija published the *Castilian Grammar*, the first grammar of any European language. Curiously, Nebrija's *Grammar* was published in the same year as the Catholic monarchs' soldiers finally took the Alhambra, the last Moorish fortress, thus ending almost eight centuries of Moorish domination of the Iberian Peninsula—a domination initiated in the year 711 by the African general Tarik in Gibraltar and later supported by another African general, Júsuf.

The Moorish conquest had enriched the Spanish language with Arabic words, but it was in the Caribbean that Spanish quickly accumulated indigenous Caribbean and African words. Referring to the enrichment of the Spanish language in the Caribbean, Jorge E. Porras writes:

> Spanish is believed not to exhibit significant substractal influence from Indoamerican or African languages in its phonological, morphological, or syntactic components but it certainly exhibits much of an influence in its lexicon. Just as in Medieval times, when Castilian [borrowed] words from other languages Latin American Spanish enriched its lexical stock with Native American languages such as Arawak (Taíno), Nahuatl, Mayan, Quechua, Tupí-Guaraní, Mapuche, and from African languages such as Kikongo, Kishigongo, Kimbundu, Ewe, and Yoruba. (1993, 181)

Some Arawakan, like canoe, hammock, cacique, bohio, tebooron, barbacoah, batata, hurricane, and maize, are now part of Latin American Spanish, as the result of linguistic syncretism, or mixing. Equally important are some examples of Africanisms: bomba, babalap, bilongo, bongó, conga, cumbia, chéchere, gahngah, guandú, geenay, lucumí, malembe, mambí, marimba, motete, ñame, ñinga, samba, tumba (see Mosonyi 1993; also see Megenny 1993).

The mixing of indigenous, European, and African languages in Caribbean Spanish, and other developments such as the Palenquero spoken in San Basilio de Palenque in Colombia, the Palenquero spoken among the descendants of maroon runaway Congo slaves of Portobelo in Panama, and the Papiamento spoken in Curaçao, are very much like the religious syncretisms of Santería, Regla, Abakuá, Palo Mayombé, Ñañiguismo, Baquiné, Voodoo, Macumba, Candomblé, as well as the musical syncretisms of Rumba, Samba, Plena, Bomba, Mambo, Bamba, Huapango, Bamboula, Cumacos, Chibángueles, Quichimba, Carángano, Quitiplás, Tango, Milonga, Beguine, Merengue, Cumbia, and Tamborito. They are also like the mixing that takes place in food: rice with chick peas, ajiaco, mondongo with lima beans.

This important syncretism or mixing of languages, religions, music, and cuisine are all original developments of the marvelous Caribbean cradle-hammock, but one of the most extraordinary telluric developments of Caribbean syncretism is the Garifuna culture. According to the research of direct descendants of the Garifuna people, this culture dates back to the mid-seventeenth century, when a hurricane in the Caribbean Sea caused slave ships coming from Africa to crash on the coast of the island of Yurumei, known today as Saint Vincent, near the coast of Venezuela. The shipwrecked Africans' odyssey had started with the abduction of children stolen from their cradles and forced to board the slave ships anchored off the Atlantic coast of Africa. Those who survived this odyssey shared food and huts with the Arawakans and Caribs when they alighted on the island of Yurumei.

There, the African children learned to communicate with their neighbors in a language that was a curious mixture of two indigenous languages: Arawakan and Carib. This Arawakan-Carib syncretism was born when the bellicose Caribs invaded the Caribbean islands and sentenced to death the Arawakan men on the island of Yurumei. The descendants of Carib fathers and Arawakan mothers taught Carib to their male offspring and Arawakan to their female offspring. In the manner of their Arawakan Carib neighbors, the shipwrecked Africans progressively became Garifunas in an interesting process of syncretism, in which they not only adopted the basic staples of the Indians on the island of Yurumei—yucca and cassava bread—but also contributed to the Arawakan-Carib language with French, English, and Spanish words (the result of their contact with Africans and slave traders, pirates, corsairs, and *flibustiers*). Above all, the intonation of African languages also influenced the Arawakan-Carib language, which later became the Garifuna language.

Scholars of Garifuna culture have pointed out that the British settlers in Yurumei—in an attempt to match the prosperity that the French settlers of Haiti gained from their fruitful sugar cane fields—tried to enslave the Garifunas who had, thanks to their shipwreck, escaped the yoke of slavery. When these settlers tried to capture the Garifunas—as free labor for the sugar cane fields, and mills for the production of highly coveted sugar, the Garifunas launched a revolt under the leadership of Satuyé—the greatest Garifuna hero. When the Garifuna warriors—armed primarily with machetes—were defeated by the firearms of the British, as punishment for defending their dignity and rejecting the yoke of slavery, Satuyé was executed on March 14, 1795. Five thousand Garifuna followers of Satuyé were captured in the Palenques (Indian ranches) of the island of Yurumei. Two years later, on March 11, 1797, the Garifunas who didn't accept enslavement were deported in eleven ships headed for Jamaica. More than a thousand of them died aboard the British ships, and after painful sailing in the Caribbean, the surviving Garifunas were abandoned by the captains on April 12, 1797, on the island of Roatán near the coast of Honduras. In Honduras the Arawakan-Carib-African syncretism or mixing continued, and there the Garifunas established their capital in Trujillo. Later they established Garifuna settlements in the Caribbean coast of Honduras: Ciriboya, Carozal, Sambuco, San Juan, Tornabé, Triunfo de la Cruz, Saraguaina, Masca, and other communities. Garifuna communities were also established in Livingston, Guatemala; in Orinoco and La Fe in Nicaragua; and in Stann Creek, Hopkins, Dangriga, and Punta Gorda in Belize.[1]

The Garifuna language- is the main patrimony of the Garifuna culture. And this Arawakan-Carib-African linguistic syncretism demands attention because although the phonetics of the language are African, unlike other Palenquero languages (such as the Palenquero spoken in San Basilio de Palenque in Colombia or the Lucumí which is sung in the ceremonies of Santería in Cuba and Candomblé in Brazil), the base of its vocabulary is not African. According to Professor Salvador Suazo,

> The linguistic structure of the Garifuna language is made up of 45 percent Arawakan words, 25 percent Kallina or Carib words, 15 percent French words, and 10 percent English words. The remaining 5 percent is made up of technical Spanish words [for the Garifunas-speakers in Honduras, Guatemala, and Nicaragua] and of English [in the Garifunas communities of Belize and among residents of the United States of America]. (Suazo 1991, 6)

Suazo offers some examples of the Carib contributions to the Garifuna language: wuguri (man), wuri (woman), arutubu (hammock), yagana (canoe), fágayu (oar); Gallicisms: weru (verre), músue (mouchoir), gulíeri (cuiller), búnedu (bonnet), mariei (maríe); Anglicisms: súgara (sugar), wachi (watch), machi (matches), haiwata (high water), giali (girl).

Garifuna culture is an important development of Caribbean syncretism which can counter both the images generated through a colonial educational system and—more devastatingly—through the popular images that we, in some nations of Central America and the Caribbean, have of Africa and of Africans in the New World. These popular images are still those of the films of Tarzan, King of the Jungle and Great Savior, who in Africa constantly defeats the dangerous Africans (the majority of whom are Pygmies and cannibals) who supposedly were the ancestors of the slaves in the Caribbean. Unfortunately, our public-school textbooks continue to present the African aspects of our culture and history through an emphasis on the African slaves who were happy because Christianity saved or delivered them from pagan and dangerous Africa. Emphasis is also placed on the "ungrateful" African slaves, who, instead of loving their masters for the salvation of their souls, dedicated themselves to marooning activities or to fighting the yoke of slavery, rescuing their human dignity, and obtaining their freedom. Thus, our students never learn from their textbooks that Africans participated in the great Pharaonic civilization in Egypt, as well as in other rich and powerful kingdoms in Nubia, Ethiopia, Mali, Shongay, Ghana, and Zimbabwe. Nor do they learn from official textbooks about the heroic deeds of conquistadors, explorers, and maroon chiefs of African ancestry such as Juan Garrido, Nuflo Olano, Juan Valiente, Estebanico, Yanga, Bayano, Ganga Zumba, Benkos, Satuyé, Fillipa Maria Aranha, Fabulé, Chirinos, Coba, Felipillo, and José Antonio Aponte (see Rout 1976).

Garifuna culture is indeed an outstandingly positive example of Caribbean syncretism, underscoring the pride and courage of Africans in the New World who rejected the yoke of slavery (and, when necessary, defended their dignity with their lives), but this syncretism has also been a painful cornucopia.

Painful are the almost four centuries of African slavery in the New World. The yoke of African slavery in the Caribbean began in 1517, when Bartolomé de las Casas petitioned Carlos V to concede licenses to Spanish settlers to import to Santo Domingo black slaves directly from Africa, a solution to the genocide of the Indians, and as a substitute for Indian slave labor.[2]

The unquestionable and undeniably important African contributions to the Creole cultures of the Spanish Caribbean stand out in Santería, in Ñañiguismo, in Merengue, in Bembé, in Bongó, in Rumba, in Ajiaco, and in the Garifunas, amongst others. However, as far as a Hispanic Caribbean identity is concerned, the African heritage is not only rejected (as in the obsessive preoccupation with racist sayings such as: "the race must be improved" and "your grandmother, where is she?"), it is also denied. Aside from being a fanatic illusion, this obsession is also an example of the profound and hateful racism that many Cubans show when they affirm that the "true Cuban" is white; that many Puerto Ricans exhibit when they proudly proclaim that Puerto Rico is "the whitest island" of the Antilles; that many Dominicans demonstrate when they swear to be "Dark Indians" and not black like the Haitians; and that many Panamanians manifest in their passionate hatred of Chombos.

In Panama the best example of the negative consequence of creolization is the separation and national hatred that exists among the so-called colonial blacks (descendants of African slaves dating back to Vasco Núñez de Balboa) and the black West Indians (disrespectfully called "Chombos"). This latter group is composed of the descendants of two waves of, English- and French-speaking West Indian workers from Barbados, Haiti, Grenada, Jamaica, Martinique, St. Lucia, and other islands. The first wave emigrated in 1850 to participate in the construction of the trans-Atlantic railroad—a project financed by the North Americans during the California gold rush. The second wave of West Indian workers emigrated to participate in the construction of the failed sea-level canal (under the direction of the French), as well as in the construction of the lock-canal, 1904-1914 (under the direction of the North Americans).

Many Panamanians hate the Chombos because they are not all Catholics (since their grandparents were originally from the West Indies, many of them practice other religions); because they prefer to speak French and English in their homes; and finally because, according to racist Panamanians, too many Chombos have failed to participate sufficiently in the process of ethnic whitening in order to "better the race"—or to put it more frankly, to erase all that is African. As a result, all traces of an African gene or phenotype is hated and rejected: the woolen hair; the flat, broad nose; the thick lips; and above all, the black skin of the Chombos.[3]

The Cuban poet Gabriel de la Concepción Valde's, Plácido, was one of the first in the Caribbean to denounce the racist obsession with whitening the race:

> Don Longuino always claims
> With a passion stronger than bacon skin,
> and with his sallow complexion
> which African lineage betrays,
> "I come from pure and noble blood."
> Deluded, he proclaims to be
> from sublime kinship!
> Let him tell it to his grandmother!
>
> (Castellanos 1984, 48)

On the island where many are proud to be natives of "the whitest island" of the Caribbean, the Puerto Rican writer José Luis González has stated:

> As far as the African roots of Puerto Rico popular culture are concerned, I am convinced that the essential racism of the island's ruling class has done everything possible—at times in brutal ways and at times with subtlety worthy of a better cause—to avoid, to conceal, and to distort its importance. (1989, 74)

This Cuban, Puerto Rican, and Dominican obsession with whitening has been synthesized and has become in Panama the cornerstone of both the "Panameñista" concept and the Constitution of 1941. The latter solely welcomes those immigrants who are "capable of contributing to the improvement of the race" and calls for the denationalization of the Panamanian [Chombos] descendants of grandparents and parents of illegal immigration who are "members of the black race whose original language was not Spanish" (*Constitution de la Republica de Panama*, 5-7).

In my own essays, poems, short stories, and novels, I have denounced and condemned the aspects of creolization that have as their sole goal and intention to erase the African heritage in Caribbean culture and identity.[4] In other words, I denounce the rejection of the African in the process of creolization which initially began with the rape of young African slave girls and which still persists today in the hatred concealed in the edict: "It is necessary to better or improve the race." For example, in *Chombo,* I display that rejection:

Abena Mansa Adesimbo vehemently opposed the name that had been given to her child. She argued that they should forget the African traditions because it was important to keep in mind that they weren't in Haiti, Jamaica, Barbados, and even less, in Africa. She recalled that during the short time that she was able to attend public school, the teacher severely criticized her name for being so African and asked her daily why West-Indian blacks did not adopt Panamanian last names such as Chiari, Wong, Heurtemate, Ghandi, Tagaropoulos. (1981, 59)

In a poem entitled "In Exilium" (1977, 8), I protest against any form of syncretism that has as its only intent the erasure and destruction of the heritage and pride of my African ancestors:

How disgraceful!
I am ASHANTI
and they address me as
Carlos
How insulting!
I am a CONGOLESE
and they call me
Guillermo
How base!
I am YORUBA
and they name me
Wilson.

Another poem, "Desarraigado" (or "Uprooted") articulates the psychological conflict produced by the erasure of the African heritage in the Caribbean identity:

African grandmother,
Do you not recognize me?
My language is Gongoric.
My litany is Nazarene.
My dance is Andalusian.
African grandmother,
Why don't you recognize me?

Finally, in the novel *Los nietos de Felicidad Dolores* (1991) I also portray characters of African descent who absurdly surrender to and become accomplices of the ideology of whitening:

Blaaaaack woman of the devil. I have already told you a thousand times not to get involved in my business. If I want to give my telephone number to all the American soldiers, of course, to the whitest with blue eyes, it's my business and au contraire, it should not matter to you nor to anyone else. And don't remind me that I have five illegitimate petits enfants because I don't feel ashamed of it. In fact, I am very happy and, yes, very proud that there were five blond soldiers, yes, very blond with blue eyes, the ones that made me pregnant. All of them white.

Well, I did as my Godmother Karafula Barrescoba advised me: "Look for a white husband in order to improve the race." Fortunately, my five children don't have woolen hair like those chombo boys with so much African blood. Neither are they thick lipped. Nor snub nosed. Neither are they... (75)

Creolization was indeed the inevitable result of the initial violent clash among the Indians who lived in such places as Quisqueya, Xaymaca, Borinquén, and Cuba; the European conquistadors who invaded these territories of the Caribbean Sea; and the African slaves brought to the New World to excavate gold mines, cultivate sugar cane fields, work in sugar mills, and build fortresses and ports. The ensuing mixing of the languages, religions, music, and food was indeed positive and admirable. Sadly, this creolization has not defeated the absurd, repugnant, and above all, insulting attitude that is an affront to the pride and dignity of the African heritage of our Caribbean cultures.

Notes

1.   This information is gleaned from the following works of scholars of Garifuna culture: Lopez Garcia, Suazo, Savaranga, Avila; I also refer to the papers of Melecio R. Gonzales and Jorge Bernardez in the Primero and Segundo Encuentro Cumbre Garifuna, held in New York in 1991 and Los Angeles in 1992, respectively.

2.   As a Sevillian soldier, Las Casas had participated in the killings and conquests of the Indians in Quisqueya and Cuba; he later became the first ordained Dominican Catholic priest in the New World, the Apostle of the Indians, and later the Bishop of Chiapas (see Sauer 1984).

3. For more details of this practice, see Jackson 1988 and Birmingham-Pokorny 1993.

4. The works I refer to here are my novels *Chombo* (1981) and *Los nietos de Felicidad Dolores* (1991) and my poems "In Exilium," and "Desarraigado" (*Pensamientos* 1977).

REFERENCES

Avila, José Francisco, ed. 1991. *U.SA. Garifuna.* Allen, TX: Avila.

Birmingham-Pokorny, Elba, ed. 1993. *Denouncement and Reaffirmation of Afro-Hispanic Identity in Carlos Gmllermo Wilson's Works.* Miami: Ediciones Universal.

Castellanos, Jorge. 1984. *Placido, Poeta Social y Politico.* Miami: Ediciones Universal.

González, José Luis. 1989. *El Pais de quatro pisos y otros ensayos.* Rio Piedras, Puerto Rico: Ediciones Huracan.

Jackson, Richard L, 1988. *Black Literature and Humanism in Latin America.* Athens: University of Georgia Press.

Lopez Garcia, Victor Virgilio. 1991. *Lamumehan Garifuna: Clamoo Garifuna.* Tela, Honduras: Tornabé.

Megenny, William W. 1993. "Common Words of African Origin Used in Latin America." *Hispania* 66 (March): 1-10.

Mosonyi, Esteban Emilio. 1993. "Nuestro Iegado Linguistico Africano." *Africamérica* 1(1) (January): 22-26.

Porras, Jorge E- 1993. "The Spanish Language in the Americas 500 Years After: Unity within Diversity." *Diaspora: Journal of the Annual Afro-Hispanic Literature and Culture Conference* 2(2) (Spring).

Rout, Leslie B. Jr. 1976. *The African Experience in Spanish America.* Cambridge: Cambridge University Press.

Sauer, Carlos Ortwin. 1984. *Descubrimiento y dominacion Espanola de la Caribe.* Trans. S. Mastrangelo. Mexico: Fondo de Cultura Economica.

Savaranga, Crisanto Uayujuru. 1992. "Conferencia de la cosmovision historica, cultural del pueblo kaliponan (garifuna) kalinagus (garinagus) de Honduras." Tegucigalpa, Honduras.

Suazo, Salvador. 1991. *Conversemos en Garifuna.* Tegucigalpa, Honduras: Editorial Guaymuras.

Wilson, Carlos G. 1977. *Pensamientos.* Los Angeles.

—1981. *Chombo.* Miami, FL: Ediciones Universal.

—1991. *Los nietos de Felicidad Dolores.* Miami, FL: Edidones Universal.

CHAPTER THREE

North Atlantic Fictions: Global Transformations, 1492-1945

Michel-Rolph Trouillot

The world became global in the sixteenth century. Europe became Europe in part by severing itself from what lay south of the Mediterranean, but in part also through a Westward move that made the Atlantic the center of the first planetary empires. As such empires overlapped or succeeded one another within the modern world system, they brought populations from all continents closer in time and space. The rise of the West, the conquest of the Americas, plantation slavery, the Industrial Revolution, and the population flows of the nineteenth century can be summarized as "a first moment of globality" an Atlantic moment culminating in U.S. hegemony after World War II.

So couched, this Atlantic moment encompasses five centuries of world history and the shrinking of huge continental masses, including Asia. The designation does not refer to a static space but to a locus of a momentum. The global flows of that era were not restricted geographically to societies bordering the Atlantic Ocean. Spain's conquest of the Philippines, the British conquest of India, and U.S. control of Korea all pertain to this moment. It is no accident that such non-Atlantic ventures took place often enough when the power that launched them claimed partial or total control of the Atlantic Ocean. In short, it is the continuous centrality of the Atlantic as the revolving door of major global flows over four centuries that allows us to speak of a single moment.

Our contemporary arrogance, which overplays the uniqueness of our times, may blind us to the dimensions of what happened before we were born. It may therefore be useful to document the density, speed, and impact of the global flows that made up this Atlantic moment. I will emphasize the earliest centuries for two reasons. First, we are less likely to realize now the importance of these early flows. Second, the evidence shows that the momentum of change was planetary from the start.

THE BEGINNING OF PLANETARY FLOWS

In 1493 when Columbus returned to the Caribbean island he had named Hispaniola, he was on a different mission than on his first trip. On the deck and in the cargo space of his 17 ships were not only the instruments of conquest that he carried the first time, but also loads of crops, fruits, seeds, and animals, from sheep, pigs, goats, cattle, and chicken to onions, radishes, chick peas, wheat seeds, and grapevine plants. If the image evokes a colonial Noah's Ark, it is in part because Columbus had purposes somewhat similar to those of the biblical patriarch: He carried these crops and animals for future reproduction in the Antilles (Davies 1991:153-57; Watts 1987:90). Given the tropical climate of the Caribbean, it seems fanciful now that Columbus envisioned the possibility of growing wheat or making wine in what is now Haiti and the Dominican Republic. Yet we need to remember that his successors succeeded quite well in winemaking half a century later in unexpected places such as Chile and California. In that sense, Columbus's second trip prefigured the massive movements of goods, crops, animals, and commodities that contributed to the Atlantic moment of globality.

Novel also in that second trip was Columbus's certainty that he or others would be able to travel back and forth between the Old and the New World. The contents of his ships were premised on the continuity of planetary flows, both those he wished for and those that he could not predict. They were premised on the fact that others from Christendom would follow his steps. Among the 1,500 men on board that second voyage, in addition to the mandatory soldiers there were also specialists in farming, irrigation, and road building, whose presence presumed back and forth movement between Spain and the Caribbean. Not long after, the Castilian invasion of the American mainland signaled the true beginning of the planetary population flows.

The twentieth century did not invent mass migration. Since the seventeenth century human beings traveled *en masse* to faraway lands for much the same reasons as they do today. Except for the early gold rush of the sixteenth century, the major migrations of the Atlantic moment, voluntary or coerced, were generated by the global distribution of labor in the capitalist world system. They included the 12 million enslaved Africans brought to feed the plantation machinery during the three long centuries that the slave trade lasted, and the Europeans—mainly British, French, Spanish, Dutch, Portuguese, or Danish—and white Americans who bought the slaves in Africa, transported them to Europe, or sold and used them in the Americas. After the end of Caribbean slavery, half a million Asians were brought to the area to replenish the labor force. Most came from the Indian subcontinent and went to Trinidad or British Guiana. Yet the areas of origins span the whole of Asia from Japan to Java and Sri Lanka. Receiving areas spanned the Antillean archipelago all the way to mainland territories such as British Guiana and Suriname. At about the same time, more than 300,000 Chinese were moved to Peru, Mexico, and Cuba, which alone took more than two-thirds of the total. By 1927 Chinese were the largest racial minority in Mexico City. During that same era, thousands of Chinese and Japanese also came to the United States and thousands of Indians moved to East Africa. In the 1920s and 1930s hundreds of thousands of Japanese moved to Brazil and Peru, a flow that did not stop until the 1970s in the latter country.

As North Atlantic states forcibly moved populations all over the world, their own citizens also moved from one continent to another, often between areas with temperate climates. Here again, labor was the main force behind these movements—except that it was rarely coerced. European migration increased tremendously as the nineteenth century came to a close. Between 1846 and 1924, 40 to 50 million European citizens migrated to the Americas. The vagueness of the estimate is itself an indication of the inability of the states involved to control or even measure these flows. We know that most of these migrants ended up in the United States, but millions also went to Canada, Brazil, and Argentina. By 1895, 74 percent of the population in and around Buenos Aires was foreign-born. By 1914, half of the population of Argentina was composed of foreign-born residents and their offspring. No city of that size in the United States, then or now, had such a high proportion of immigrants. Indeed, at the beginning of the twentieth century Argentina, Canada, Australia, and New Zealand had a larger immigration ratio than the United States did either then or now. The coinage of the United States as a nation of immigrants notwithstanding, European migration targeted lands outside of the Americas. Australia, New Zealand, and Southern Africa still experience the marks of these early global flows.

As peoples moved so did goods. Massive flows of gold and silver, of crops and spices, of plants and diseases, from tobacco to coconuts, from syphilis to smallpox, and from the mines of Peru to the botanic gardens sprinkled throughout the British Empire, enmeshed world populations into encounters and confrontations unrestricted by physical distance. From the beginning Europeans who came to the New World brought along with their slaves a variety of plants, animals, and other living organisms. Horses, pigs, sheep, dogs, chickens, donkeys, cattle, bananas, plantains, and all their parasites moved to the New World. So did measles, whopping cough, bubonic plague, malaria, yellow fever, diphtheria, amoebic dysentery, influenza, and smallpox. The later alone proved to be a mass murderer of proportions still unmatched for the native population.

While the movement of peoples and animals between the Old World and the New was largely unidirectional, that of crops was not. American crops that spread into Europe, Asia, and Africa included maize, potatoes, tomatoes, peanuts, manioc, cacao (chocolate), tobacco, and many types of peppers, beans, and squashes that were unknown to Europe before the time of Columbus (McNeill 1992).

Other crops and their by-products fully completed the global circle, in the process becoming planetary commodities. Domesticated outside of Europe, they were brought to the Americas by Europeans only to be resold later to Europe or even to African or Asian clients. The first of these was sugar cane. Originally domesticated in New Guinea and introduced to Europe by way of South Asia, in the eighteenth century it became the main export of many Caribbean slave territories, vivifying the European proletariat and the North Atlantic predilection for sweetness (Mintz 1985). Similarly coffee, first domesticated in Yemen and brought to the Caribbean in the eighteenth century, was soon to be resold to Middle Eastern clients of Prance (Trouillot 1982). The British Admiralty in Africa would later use Caribbean citrus as a protection against scurvy. Bananas, an Old World cultigen, became the main export crop of the Windward Islands, Colombia, and Ecuador, Tobacco, cacao, rice, and to a lesser extent opium and manioc, also became global commodities of this Atlantic moment.

These flows of commodities sustained the life of the North Atlantic both before and after its Industrial Revolution. Economic elites had speculated on the returns from these exchanges, with varying degrees of respect for political boundaries, since at least the sixteenth century. By the seventeenth century, the wheels of exchange had planetary dimensions (Braudel 1992). The history of Holland from the 1592 reopening of the Amsterdam stock market to its crash in 1783 is a textbook story of merchant and finance capital crossing borders, linking continents, and in the process affecting local beliefs and practices. It is also a story of private enterprise dominating states in ways that we often believe unique to our own times. Dutch, merchants backed the Spanish colonial enterprise in the Americas, then backed their own fleet against Spanish and Portuguese vessels, then provided credit to France and England while turning Amsterdam into a huge depot for commodities from all continents. In so doing, they accumulated a global power unmatched by any royalty. The Dutch West India Company established warehouses in Brazil, Curaçao, and New York. The Dutch East India Company, the equivalent of a transnational powerhouse using the services of 8,000 sailors, developed a profitable trade within Asia along its own transcontinental axis of spices, selling wood from Timor to China, Indian textiles to Sumatra, and Siamese elephants to Bengal. In short, early on in the first moment of globality, capital, labor, and the commodities they generated circumscribed a world of which the various subparts were increasingly intertwined in ways that we now tend to forget.

The flow of goods and capital across political and geographical boundaries was not always increasing but it did reach a peak in the period immediately preceding World War I. Ratios of export trade to GDP may have been higher in 1913 than in 1973. In 1913-14, Foreign Direct Investment (FDI) was around 11 percent, about the same level as in 1994. Capital flows relative to output were higher during the early decades of the twentieth century than in the 1980s.

In that sense, World War I was aptly named if only because it confirmed the global ties that these figures suggest. The Great War involved the seizure of German holdings in Oceania, Southwest Africa, and Tanganyika, Indians fought on the British western front and Senegalese *tirailleurs* died in France and for France. Eleven years after the war, the great crash of the 1930s tied New York and Vienna together in a downward spiral that sent the prices of agricultural goods from all over the globe plummeting. It was indeed a world depression, soon followed by a second world war.

CHANGING PRACTICES, COMPLEX IDENTITIES

These massive movements of goods, populations, and capital produced abrupt changes not only in the material conditions of the populations involved but also in their practices and in the ways in which they saw themselves and the world around them. We tend to think of our contemporary era as one of swift transformations that challenge our capacity for adaptation, and indeed it is. Yet the first moment of globality was also characterized by speed for many of those who lived it, and it constantly tested their capacity to adapt swiftly. They passed that test more often than we now think.

Again, the sixteenth century gives us a glimpse of the global momentum-Maize most probably went to the Old World during one of Columbus's return trips. By the 1560s it was cultivated in places as distant from the Americas and far away from each other as the West Coast of Africa and the Hunan province of China. At about the same time, Spanish friars were setting up the first wineries of Chile, Peru, and California. Native Mexicans, who did not know cattle before the Conquest, were then working on ranches, some of which counted 150,000 heads of cattle.

Colonialist exploitation was often the motor behind these swift adaptations, especially in the eighteenth and nineteenth centuries when colonial control over production generally became more systematic. Yet direct colonial control was not always a factor. At times the new import provided a clear advantage on its competitors, as did the potato throughout most of Europe. Some Native Americans quickly adopted the horse from flocks that had escaped Spanish ranches. By the time they encountered the first Anglo-Saxon colonists, they had already integrated horseback riding into their daily cultural practices. It took less time for maize to be adopted by Africans on the Angola-Congo coastline, in the late sixteenth century than it took espresso to move inland from the two coasts of the United States and become an accessible commodity in the Midwest or the South in the late twentieth. One could argue that maize mush is inherently more agreeable than coffee in its espresso form, but such an argument implies a value judgment on the universal acceptability of edibles, all of which are culturally marked. On surer ground, one could demonstrate that colonial pressures and the political economy of the African coast at the time—including the cost accounting of cereal production, down to individual caloric intake—made the acceptance of maize relatively easy. Africans could swiftly adopt maize because it was practical for them to do so then. That argument immediately relativizes our sense of our own cultural openness.

Such speedy changes affected political and cultural identities and practices and provoked reactions varying from revolt to acceptance to confusion. Not surprisingly, the first moment of globality produced its self-proclaimed hybrids, individuals or groups who saw themselves as belonging to more than one sociocultural unit and as sharing more than one cultural heritage. Seventeenth-century Cambay, a commercial port on the Indian Ocean linking East Africa, the Middle East, and Indonesia, counted a number of Portuguese residents who erected mansions built and furnished according to Portuguese taste. Was it an Indian, Islamic, or Portuguese city? K. N. Chaudhuri (1990:347), who asks that question, answers: "It was all three simultaneously as an abstraction but one or the other according to the viewpoint of its inhabitants." Although Cambay had distinct ethnic quarters, one suspects that the abstraction and the quite concrete presence of each quarter impacted on each group's sense of identity. We know that it was impossible for Dahomeans in Barbados, Japanese in Peru, Javanese in Suriname, or Indians in East Africa to escape the sense of being caught between two worlds. It may have been equally hard for their children to pick any one of the two.

The sense of belonging to many worlds must have also been common among many of the *convertos* (Jews forced to convert to Christianity) who joined the Castilian venture to the Americas. Cultural overlap reinforced by power equally marked the Filipinos first brought under the Spanish umbrella, then forced by the United States to manipulate cultural streams of various densities and provenance. Also hybrids of a kind were the early Americans who discovered they had become "Indians" and were coached to write in Spanish the history of an Indianness that came with the conquistadors. Self-proclaimed hybrids were the mulattos of Cuba, Brazil, Saint-Domingue, and Louisiana, and many Latin. American mestizos. By

1815, Simon Bolivar had officialized a narrative of *mestizaje:* "We are... neither Indian nor European, but a species midway between the legitimate proprietors of this country and the Spanish usurpers." Clearly the praise of diversity and the celebration of mixed origins are not so new. Nor are their use for political gains. In Latin America as in the Caribbean, the consciousness of mixed origins has been widespread for centuries. Some authors argue that the awareness of cultural *metissage* is inherent in the creolization process as it developed in the Antilles, and thus inherent in Caribbean life. In short, since the early centuries of the Atlantic moment, identities have never been as simple as we are sometimes prone to believe.

The awareness of mixed origins does not mean that individuals can spontaneously retrace the flows that contributed to shaping their current practices and environment. Indeed, the long-term impact of cultural imports is often proportional to the capacity to forget that they were once acquired or imposed. How many Californians routinely ponder the Spanish names of their streets and towns? How many Italians today do not see the tomato as an intrinsic part of their cultural heritage? How many Native American leaders would dare to reject the horse as culturally foreign? In stressing the impact of the plants exported from the Americas to the Old World, William McNeill (1992:34-5) asks us to imagine the Italians without tomatoes, the Chinese without sweet potatoes, the Africans without maize, and the Irish, Germans, and Russians without potatoes. From the record sketched above, we could prolong the list interminably in a number of directions: Latin America without Christianity, India without English, Argentina without Germans, Texas without cattle, the Caribbean without blacks or rum, England without tea, France without cafés, or French fries. The point is obvious. Culturally, the world we inherit today is the product of global flows that started in the late fifteenth century and continue to affect human populations today. Yet the history of the world is rarely told in those terms.

Indeed, the particularity of the dominant narratives of globalization is a massive silencing of the past on a world scale, the systematic erasure of continuous and deep-felt encounters that have marked human history throughout the globe and that I have only sketched here. For sushi in Chicago to amaze us, we need to silence the fact that the Franciscans were in Japan as early as the fifteenth century. For Muslim veils in France to seem out of place, we need to forget that Charles Martel stopped Abd-al-Raman only

300 miles south of Paris two reigns before Charlemagne. To talk of a global culture today as a new phenomenon, we need to forget that Chinese chili paste comes from Mexico, French fries from Peru, and Jamaican Blue Mountain coffee from Yemen.

A central task, then, for historical anthropology is to bring to public consciousness these flows that shaped the world in which we live. Yet the vulgarization of the historical record is not enough. After all, these facts were always part of the available record. That they were rarely accorded the significance they deserve suggests the existence and deployment of mechanisms of silence that make them appear less relevant than they are, even when they are known. The silencing of the past inheres not only in *what* is said but also in *how* it is said (Trouillot 1995).

Thus, a theoretical task parallel to the documentation of these flows is to assess the terms of the dominant narratives of world history—the words used, the concepts deployed, the setting of the plots and subplots, the depiction of the characters and the connections made or ignored between all of the above. We should hold under suspicion any word that describes a chunk of the story while claiming universal relevance. Words such as progress, development, modernity, nation-state, and globalization itself are among those I have in mind. The beginning of this chapter should have raised some doubts about the abuse of the word globalization. The following sections demonstrate further how suspicion toward these master words is well founded by way of an exploration of "modernity," a term increasingly yet differently used by anthropologists (Appadurai 1996; Gaonkar 1999; Knauft: 2002a,b).

NORTH ATLANTIC UNIVERSALS

Modernity is a murky term that belongs to a family of words we may label "North Atlantic universals." I mean by that words that project the North Atlantic experience on a universal scale that they themselves have helped to create. North Atlantic universals are particulars that have gained a degree of universality, chunks of human history that have become historical standards. Words such as development, progress, democracy, and nation-state are exemplary members of that family that contracts or expands according to contexts and interlocutors. Belonging to that class does not depend on a fixed meaning. It is a matter of struggle and contest about and around these universals and the world they claim to describe. Only time will tell if newly popular expressions such as "globalization" or "the international community" will become North Atlantic universals.

North Atlantic universals so defined are not merely descriptive or referential. They do not describe the world; they offer visions of the world. They appear to refer to things as they exist, but rooted as they are in a particular history they are evocative of multiple layers of sensibilities, persuasions, cultural and ideological choices tied to that localized history. They come to us loaded with aesthetic and stylistic sensibilities, religious and philosophical persuasions, cultural assumptions ranging from what it means to be a human being to the proper relationship between humans and the natural world, ideological choices ranging from the nature of the political to its possibilities of transformation. There is no unanimity within the North Atlantic itself on any of these issues, but there is a shared history of how these issues have been and should be debated, and these words carry that history. Yet since they are projected as universals, they deny their localization, the sensibilities, and the history from which they spring.

North Atlantic universals are always prescriptive inasmuch as they always suggest, even if implicitly, a correct state of affairs: what is good, what is just, what is sublime or desirable—not only what is, but what should be. That prescription is inherent in the very projection of a historically limited experience—that of the North Atlantic—on the world stage. North Atlantic universals not only prescribe: They seduce. Indeed, they are always seductive, at times even irresistible, exactly because they manage, in that projection, to hide their specific—localized, and thus parochial—historical location. This power of seduction is further enhanced by a capacity to project affect without actually claiming to do so. All ideas come with affect, but a successful universal tends to hide the affect it projects behind a claim of rationality. It makes sense to be modern. It is good to be modern. How could anyone not want to be modern? Similarly, how could anyone not want to join the international community? To be sure, these propositions mean different things to different people. At the same time, the number of divergent voices that use and abuse these words verify their attraction. One might go as far as to say that the capacity to seduce is inherent in such universals.

Their ability to project transhistorical relevance while hiding the particularities of their marks and origins, including their affective load, makes North Atlantic universals as difficult to conceptualize as they are seductive to use. The more seductive these words become the harder it is to specify what they actually stand for, since part of the seduction resides in that capacity to project clarity while remaining ambiguous. Even if we accept the questionable assumption that concepts are merely words, a quick perusal of the popular press in any European language demonstrates that North Atlantic universals are murky references: They evoke rather than define. Furthermore, even that evocation works best in negative form. We have a stronger sense of what modernity may connote when we point to the naysayers—the Taliban of Afghanistan, a native tribe in the Amazon, or whichever figure plays temporarily the good or evil face of the non-modern— than when we investigate those who praise it. The seduction and the confusion are related. Dreams of a democratic future, practices and institutions of a democracy at work, or claims to join and to defend the international community vary in time and place. Even who actually belongs to the international community is a matter of contention, as any debate of the UN. General Assembly demonstrates. Attempts to conceptualize North Atlantic universals in the scholarly literature reveal little unanimity about their scope, let alone their denotation (Dussel 1993; Gaonkar 1999; Knauft 2002a).

Thus, I am quite ambivalent about the extent to which modernity can be fully conceptualized. At the same time, it would be disingenuous not to acknowledge that the word modernity evokes sensibilities, perceptions, choices, and states of affairs that are not easily captured *by* other words. That is in part why it is a seductive word. But if the seduction of North Atlantic universals also has to do with their power to silence their own history, then we need to unearth those silences, those conceptual and theoretical missing links that make them so attractive. Insisting on such silences, I argue that in its most common deployments as a North Atlantic universal, modernity disguises and misconstrues the many Others that it creates. A critical assessment of modernity must start with the revelation of its hidden faces.

THE MANAGEMENT OF IMAGINATION

Modernity and modernization each call to mind the necessary coexistence of the two geographies through which the deployment of the West and the deployment of world capitalism take place. As moments and aspects within these deployments, yet figures within two distinctive geographies, modernity and modernization are both discrete and intertwined. Thus, a rigid distinction between societal modernization and cultural modernity can be misleading (Gaonkar 1999:1), especially when it couches them as separate historical developments that can each be judged on its own terms. But the distinction remains useful if we keep in mind that the bundle of facts and processes we package under one label was at any moment of world history, *as a package*, a condition of possibility of the processes and phenomena that we cover with the second. The distinction becomes necessary inasmuch as it illuminates specific historical moments and processes.

To speak of modernization is to put the accent on the material and organizational features of world capitalism in specific locales. It is to speak of a geography of management, of those aspects of the development of world capitalism that reorganize space for explicitly political or economic purposes. We may note among the continuities and markers along that line the French Revolution as a moment in the modernization of the state, as a reorganization of space for political management. We may read the English Industrial Revolution as a moment in the reorganization of labor relations, here again a reorganization of space primarily for economic purposes. Similarly, the wave of decolonization following World War II can be read as a moment in the modernization of the interstate system, one more moment of reorganization of space on a world scale that provides a new geography of management. Closer to our times, what we now call globalization—and which we too often reduce to a concoction of fads and slogans—inheres in a fundamental change in the spatiality of capital (see chapter 3). In short modernization has everything to do with political economy, with a geography of management that create *places*: a place called France, a place called the Third World, a place called the market, a placed called the factory or, indeed, a work-place.

If modernization has to do with the creation of place—as a relation within a definite space—modernity has to do with the projection of that place—the local—against a spatial background that is theoretically unlimited. Modernity has to do not only with the relationship between place and space but also with the relation between place and time. In order to prefigure the theoretically unlimited space—as opposed to the space within which management occurs—one needs to relate place to time, or address a unique temporality, that is, the position of the subject located in that place.

Thus modernity has to do with those aspects and moments in the development of world capitalism that require the projection of the individual or collective, subject against both space and time. It has to do with historicity.

I will further expand on that argument both in discussing the work of Reinhart Koselleck (1985) and in discussing features of Caribbean history. For now we may note as markers of modernity historical moments that localize the individual or collective subject while opening its spatial and temporal horizons and multiplying its outside references. The invention of private life in the Renaissance — and the accompanying features noted by Roger Chartier (1989) and others such as the spread of silent reading, personal journals, private libraries, the translation of the Bible into vernacular languages, the invention of the nation and national histories, and the proclamation of the U.S. Bill of Rights, can all be read as key moments in the spread of modernity. Closer to our times, the global production of desire spurred by the unification of the world market for consumer goods expands further the geography of imagination of which modernity is part (see chapter 3).

This last example is telling. That this global production of desire as a moment of modernity parallels globalization as a moment in the spatial history — and thus the management — of capital suggests that although modernity and modernization should not be confused, they are inherently intertwined. One could take the two lists of markers that I have suggested, extend them appropriately and draw lines across them that spell out this inextricability. From the printing press to silent reading, from the political rise of the bourgeoisie to the expansion of individual rights, from the elusiveness of finance capital to the elusiveness of global desires, the geography of management and the geography of imagination are intertwined. Just as the imaginary projection of the West constantly refuels managerial projects of modernization, modernization itself is a condition of possibility of modernity.

HISTORICITY AND ALTERITY: THE MODERN AS HETEROLOGY

As part of the geography of imagination that constantly recreates the West, modernity always required an Other and an Elsewhere. It was always plural, just like the West was always plural. This plurality is inherent in modernity itself, both structurally and historically. Modernity as a structure, requires an Other, an alter, a native, indeed an alter-native. Modernity as a historical process created this alter ego, which was as modern as the West—yet otherwise modern.

If we follow the line of argument drawn from Reinhart Koselleck (1985) that modernity implies first and foremost a fundamental shift in regimes of historicity, most notably the perception of a past radically different from the present and the perception of a future that becomes both attainable (because secular) and yet indefinitely postponed (because removed from eschatology), we come to the conclusion that modernity requires a localization in space in order to position subjects within the historicity it creates. Koselleck does not reach that conclusion himself, yet those of us who claim that modernity requires a geography of imagination (see chapter 1; Glissant 1989; Mudimbe 1994) are not necessarily at odds with his analysis. As soon as one draws a single line that links past, present, and future, and yet insists on their distinctiveness, one must inevitably place actors along that line. Not everyone can be at the same point along that line; some become more advanced than others. From the viewpoint of anyone anywhere in that line, others are somewhere else, ahead or behind. Being behind suggests an elsewhere that is both inside and outside the space defined by modernity: outside to the extent that these others have not yet reached that place where judgment occurs; inside to the extent that the place they now occupy can be perceived from that other place within the line. To put it this way is to note the relation between modernity and the ideology of progress (Dussel 1993), between modernity and modernism. But there is more to the argument.

In his treatment of modernity, Koselleck insists upon historicity—that is, in part, a relation to time of which the chronologization, periodization, distanciation, increasing speed, and range of affective relations from hope to anxiety help to create a new regime. But if he is correct, as I believe he is, this new regime of historicity also requires a localization of its subject. Time here creates space. Or more precisely, Koselleck's historicity necessitates a locale, a *lieu* from which springs this relation to time. Yet by definition, the inscription of a lieu requires an Elsewhere—a space of and for the Other. That this space can be—indeed, often is—imaginary merely suggests that there may be more continuities than we think between the geography of imagination of the Renaissance and that of the Enlightenment.

Within that geography, elaborations of a state of nature in Hobbes, Locke or Rousseau, as varied as they indeed are between and across these authors, emerge as alternative modernities—places, locales against which we can read what it means to be modern. Rousseau is the clearest on this for two reasons. First, he is not a modernist. He does not believe in either the inevitability or the desirability of linear progress. Indeed, critics wrongly accuse him of naïveté vis-à-vis the noble savage and earlier stages of human history. Second, that critique notwithstanding, Rousseau explicitly posits his state of nature as a structural and theoretical necessity to which the historical reality is largely irrelevant. He needs that fictional time to mark his own space as a modern one. Later observers will be less perceptive. As the line that ties past, present, and future gets more acute and more relevant, as both the momentum behind it and the goal to which it aspires become clearer—otherwise said, as teleology replaces eschatology—from Condorcet to Kant and from Hegel to Marx, the place assigned to the Other may fall not only within the line but also *off the* line. Hegel's dismissal of Africa and Marx's residual "Asiatic" mode of production—maybe his most unthought category—are exemplars of a hierarchy of spaces created through a relation to time. Not only does progress and its advance leave some people "behind" (an Elsewhere from within) but increasing chunks of humanity fall off its course (an Elsewhere on the outside that can only be perceived from within). The temporal-historical regime that Koselleck associates with modernity creates multiple spaces for the Other.

If that is so, modernity necessitates various readings of alterity, what Michel de Certeau (1986) calls an heterology. The claim that someone—someone else—is modern is structurally and necessarily a discourse on the Other, since the intelligibility of that position—what it means to be modern—requires a relation to otherness. The modern is that subject that measures any distance from itself and redeploys it against an unlimited space of imagination. That distance inhabits the perspectival look to and from the painted subject in Raphaël or Titian's portraits. It fueled the quarrel of the Ancients and Moderns in Louis XIV's France. It is crucial to Charles Baudelaire's (re) definition of modern art and poetry as both recognition and rejection of time.

BAUDELAIRE'S SHADOW

Idiosyncratic as it may be, the case of Baudelaire suggests in miniature the range of silences that we need to uncover for a critical assessment of modernity that would throw light on its hidden faces. As is well known, Baudelaire had just turned 20 when his stepfather forced him to embark for Calcutta. He went only as far as Mauritius and Bourbon (now Réunion), then part of France's plantation empire. That trip inspired—and may have seen the first drafts of—many of the poems that would later be published in *Les Fleurs du Mal*. Back in Paris, Baudelaire entered into a relationship with a "mulatto" actress, better known as Jeanne Duval, widely said to be of Haitian descent. Although Baudelaire's liking of dark-skinned females seems to have preceded that liaison, his tumultuous affair with the woman he called his "Black Venus" lasted over 20 years, during which she was a major source of poetic inspiration for him.

Only recently has the relationship between Mme. Duval and Baudelaire become a central object of scholarly research.[1] Emmanuel Richon (1998) points out that Baudelairian scholarship has not even bothered to verify the most basic facts about Duval, including her actual origins. The many sketches of Duval by Baudelaire, and other portraits such as Edouard Manet's "La maitresse de Baudelaire couchée," only confirm her constant presence in his life. Many visitors recount entering the poet's place to find him reading his unpublished poetry to Jeanne. Literary scholarship has attributed some of Baudelaire's work to a "Jeanne Duval cycle," insisting on her role as "femme fatale" and relishing the assertion that Duval infected Baudelaire with syphilis. Richon demolishes that assertion, convincingly arguing that the opposite was more likely.

But the main lesson of Richon's work goes beyond biographical rectification. His claim that the Indian Ocean trip and especially the relationship with Duval fundamentally shaped Baudelairian aesthetics suggests that Baudelairian scholarship may have produced what I call a "silence of significance" through a procedure of banalization. Well-known facts are recounted in passing, yet kept in the background of the main narrative or accorded little significance because they "obviously" do not matter (Trouillot 1995). Yet can it not matter that Baudelaire was living a racial taboo in the midst of a Paris sizzling with arguments for and against the abolition of slavery and the equality of human races? Slavery was abolished in Bourbon and other French possessions less than seven years after Baudelaire had been there and while he was enthralled in his relationship with Jeanne. Can it not matter that the eulogist of modernity was also Jeanne Duval's eulogist?

The issue is even more mtriguing in light of Baudelaire's own disdain for the modernization—the concrete management of places and populations by the French state, republican and imperial as it was—that was a condition of possibility of his own modernity. As in Rousseau, Baudelaire's relation to time, a hallmark of his modernity, does not imply a blind faith in either the desirability or the inevitability of progress. Indeed, Baudelaire is resolutely anti-modernist (Froidevaux 1989). His modernity is founded upon the search for a furtive yet eternal present. The past has no legacy; the future holds no promises. Only the present is alive. With Baudelaire, we are distant from either side of the quarrel between the Ancients and the Moderns and from Koselleck's regime of historicity. Baudelaire's modernity is indeed a new brand that prefigures the postmodern.

How interesting, then, that this new brand of modernity also leads to "the spatialization of time" (Froidevaux 1989:125). Baudelaire's escape from chronological temporality is space—more specifically, the space of the Elsewhere. Here again, time creates space, and here again space generates a heterology. Literary scholars have long noted the importance of themes and metaphors of space and of travel, as well as the role of exoticism, in Baudelaire's poetry. While we should leave to specialists the task of mapping out the many locations in a geography of imagination that links space and time, the Here and the Elsewhere, routine and exoticism, we may want to provoke them to find out the extent to which the modernity of Baudelaire, the critic, establishes itself against the background of an ethereal Elsewhere that Baudelaire, the poet, inscribes somewhere between Jeanne's body and the islands of the Indian Ocean.

DIFFERENTLY MODERN: THE CARIBBEAN AS ALTER-NATIVE

I have argued so far that modernity is structurally plural inasmuch as it requires a heterology, an Other outside of itself. I would like to argue now that the modern is also historically plural because it always requires an Other from within, an otherwise modern created between the jaws of modernity and modernization. That plurality is best perceived if we keep modernity and modernization as distinct yet related groups of phenomena with the understanding that the power unleashed through modernization is a condition of possibility of modernity itself. I will draw on the sociohistorical experience of the Caribbean region to make that point.

Eric Wolf once wrote in passing, but with his usual depth, that the Caribbean is "eminently a world area in which modernity first deployed its powers and simultaneously revealed the contradictions that give it birth." Wolf's words echo the work of Sidney W. Mintz (1966, 1971b, 1978,1983,1996,1998) who has long insisted that the Caribbean has been modern since its early incorporation into various North Atlantic empires. Teasing out Wolf's comments and drawing from Mintz's work, I want to sketch some of the contradictions from the Caribbean record to flesh out a composite picture of what I mean by the Otherwise Modern.

Behold the sugar islands from the peak of Barbados's career to Cuba's lead in the relay race—after Jamaica and Saint-Domingue, from roughly the 1690s to the 1860s. At first glance, Caribbean labor relations under slavery offer an image of homogenizing power. Slaves were interchangeable, especially in the sugar fields that consumed most of the labor force, victims of the most "depersonalizing" side of modernization (Mintz 1966). Yet as we look closer, a few figures emerge that suggest the limits of that homogeneity. Chief among them is the slave striker, who helped decide when the boiling of the cane juices had reached the exact point when the liquid could be transferred from one vessel to the next.[2] Some planters tried to identify that moment by using complex thermometers. But since the right moment depended on temperature, the intensity of the fire, the viscosity of the juice, the quality of the original cane, and its state at the time of cutting, other planters thought that a good striker was much more valuable than the most complex technology. The slave who acquired such skills would be labeled or sold as "a striker." Away from the sugar cane, especially on the smaller estates that produced coffee, work was often distributed by task, allowing individual slaves at times to exceed their quota and gain additional remuneration.

The point is not that plantation slavery allowed individual slaves much room to maneuver in the labor process: it did not. Nor is the point to conjure images of sublime resistance. Rather, Caribbean history gives us various glimpses at the production of a modern self—a self producing itself through a particular relation to material production, even under the harshest possible conditions. For better *and* for worse, a sugar striker was a modern identity, just as was being a slave violinist, a slave baker or a slave midwife (Abrahams 1992:126-30; Debien 1974; Higman 1984).

That modern self takes firmer contours when we consider the provision grounds of slavery. Mintz (1978) has long insisted on the sociocultural relevance of these provision grounds, small plots on the margins of the plantations, land unfit for major export crops in which slaves were allowed to grow their own crops and raise animals. Given the high price of imported food, the availability of unused lands, and the fact that slaves worked on these plots in their own free time, these provision grounds were in fact an indirect subsidy to the masters, lessening their participation in the reproduction of the labor force.

Yet Mintz and others—including myself—have noted that what started as an economic bonus for planters turned out to be a field of opportunities for individual slaves. I will not repeat all those arguments here (Trouillot 1988, 1996, 1998). Through provision grounds, slaves learned the management of capital and the planning of family production for individual purposes. How much to plant of a particular food crop and where, how much of the surplus to sell in the local market, and what to do with the profit involved decisions that required an assessment of each individual's placement within the household. The provision grounds can be read not only as material fields used to enhance slaves' physical and legal conditions—including at the time the purchase of one's freedom—they can also be read as symbolic fields for the production of individual selves by way of the production of material goods.

Such individual purposes often found their realization in colonial slave markets, where slaves—especially female slaves—traded their goods for the cash that would turn them into consumers. One can only guess at the number of decisions that went into these practices, how they fed into a slave's habitus, or how they impacted on gender roles then and now in the Caribbean. Individual purposes also realized themselves through patterns of consumption, from the elaborate dresses of mulatto women, to the unique foulard (headscarf) meant to distinguish one slave woman from another. The number of ordinances regulating the clothing of nonwhites, both free and enslaved, throughout the Caribbean in the days of slavery is simply amazing. Their degree of details—e.g., "with no silk, gilding, ornamentation or lace unless these latter be of very low value" (Fouchard 1981 [1972] :43) is equally stunning. Yet stunning also was the tenacity of slaves who circumvented these regulations and used clothing as an individual signature.

Moreau de St-Méry, the most acute observer of Saint-Domingue's daily life, writes:

> It is hard to believe the height to which a slave woman's expenses might rise... In a number of work gangs the same slave who wielded tools or swung the hoe during the whole week dresses up to attend church on Sunday or to go to market; only with difficulty would they be recognized under their fancy garb. The metamorphosis is even more dramatic in the slave woman who has donned a muslin skirt and Paliacate or Madras kerchief... (in Fouchard 1981 [1972] :47).

Moreau's remarks echo numerous observations by visitors and residents of the Americas throughout slavery's long career.

If modernity is also the production of individual selves through patterns of production and consumption, Caribbean slaves were modern, having internalized ideals of individual betterment through work, ownership, and personal identification with particular commodities. It was a strained and harsh modernity, to be sure. Otherwise modern they were; yet still undoubtedly modern by that definition.

One could argue—although the argument is not as easy as it seems—that the selves on which I just insisted may have existed elsewhere without the forced modernization imposed by colonialism. I would readily concede that point if it leads to the realization that the modern individual self claimed by North Atlantic consciousness is not unique to the North Atlantic. At the opposite extreme, one could. also argue that the detached individual self is only a fiction of the North Atlantic geography of imagination, an ideological by-product of the internal narrative of modernity. Perhaps surprisingly, I am even more willing to concede that point. In either case, the central issue is not that of an allegedly modern individual subjectivity—whatever that may be—but the insertion of that subjectivity into a particular regime of historicity. Clothing as individual signature may be as old as human society. So too may be the production of identity through labor. At any rate, I doubt that these two features—or any of the markers usually claimed to signify the rise of the modern self—first obtained as such in Renaissance or post-Renaissance Christendom. Intellectual and art history, literature and philosophy may have misled us into overrating these individual attributes of the modern self to the detriment of the historical context within which these selves were fashioned. François Hartog (1988 [1980]) sets the projection of alterity as the context for self-identification as far back as Herodotus. Max Horkheimer and Theodor Adorno see in Odysseus the precursor of the modern subject. Closer to the ground, Georges Duby and his collaborators in the *History of Private Life* project (1988) effectively extend notions of privacy or even intimacy back into the Middle Ages. I suspect that with similar data one could make as potent discoveries outside of Christendom, thus relativitizing the narrative that makes the modern individual self such a Eurocentric product.

Necessary as this revisionist narrative is, it is not the central issue. Too often critics of Eurocentrism flesh out their arguments in terms of chronological primacy. They spend much energy demonstrating that such-and-such feature claimed by North Atlantic narratives to have been a European first could actually be found elsewhere before European presence. The mistake here is to forget that chronological primacy is itself a central tenet of North Atlantic imagination. That is, the value of being the first comes from a particular premium on time, a specific take on historicity. The existence of certain social features outside of Europe matters less than the inscription of these features in social and political regimes *in the past*, and much less even than the inscriptions of these same features—as found in Europe then—in North Atlantic narratives *in the present*. From that perspective, the modern self may be less a matter of the content of an individual subjectivity than that of the insertion of that subjectivity into a particular regime of historicity and sociopolitical management. On that latter issue, the most crucial one in my view, the Caribbean story is most revealing.

Modern historicity hinges upon both a fundamental rupture between past, present, and future—as distinct temporal planes—and their relinking along a singular line that allows for continuity. I have argued that this regime of historicity in turn implies a heterology, a necessary reading of alterity. Striking then is the fact that Caribbean history as we know it starts with an abrupt rupture between past and present—for Europeans, for Native Americans, and for enslaved Africans. In no way could the enforced modernization imposed by colonization be perceived by any of these actors as a mere continuation of an immediate past. This was a New World for all involved, even for those who had lived within it before it became new to others.

The consciousness that times had changed, that things were falling apart and coming together in new ways, was both inescapable and yet inseparable from the awareness that others were fundamentally different—different in where they came from, the positions they occupied along any of the intersecting hierarchies, the languages they spoke, the costumes they wore, the customs they inhabited, and the possible futures they could envision. The sensibility to time and the recognition of heterogeneity associated with modernity are inescapable here. Indeed, they have been central themes of Caribbean scholarship (Lewis 1983; Trouillot 1992, 2001b).

Here again the slave quarters are telling. These imposed the sudden discovery of a common African past, but also the awareness that this commonality barely covered fundamental differences. One could not address that Other next door who looked so strikingly similar and engaged at times in practices reminiscent of home, without using a language derived at least in part from that of the masters. Was that not as modern as the vulgate version of the Bible? More modern than the quarrel between seventeenth-century French intellectuals as to whether the King's engravings were best written in French or Latin? If the awareness of one's position in history, not just as an individual but as part of a group and against the background of a social system brought to consciousness, is a fundamental part of what it means to be modern, then the Caribbean was modern from day one, from the very day colonialism imposed its modernization. If the awareness of sociocultural differences and the need to negotiate across such differences are part of what we call modernity, then the Caribbean was modern since at least the sixteenth century—from day one of North Atlantic modernity. But if that is so, the chronological primacy of the North Atlantic falters.

Chronology here is only an index. My goal is not to replace North Atlantic chronological primacy over the rest of the world with a Caribbean chronological primacy over other colonies and postcolonies. Historical particulars made the Caribbean, for better and for worse, the area longest under European control outside of Europe itself and the only one where Europeans moved as if it was indeed empty land, a *terra nullius* to be fashioned along modern lines. Now dominant North Atlantic narratives — reflecting the world domination of the English language, the expansion of Protestantism as a variant of Christianity, and the spread of Anglo-Saxon and Teutonic sensibilities — reduce the crucial role of Portugal and Spain in the creation of the West. A related emphasis on the Enlightenment and the nineteenth century, and the downplay of the Renaissance as a founding moment, also lead to the neglect of the role of the Caribbean and Latin America in the production of the earliest tropes associated with modernity. That chronological amnesia crucially impedes our understanding of the North Atlantic itself (see chapter 1; Dussel 1993; Trouillot 1991,1995).

Yet I want to insist that the lessons learned from the Caribbean are applicable elsewhere.. As a historical process inherently tied to modernization, modernity necessarily creates its alter-native in Asia, Africa, Latin America, and in all areas of the world where the archetypal Caribbean story repeats itself with variations on the theme of destruction and creolization. Modernity creates its Others — multiple, multi-faced, multi-layered. It has done so from day one: *we* have always been modern, differently modern, contradictorily modern, otherwise modern, yet undoubtedly modern.

I do not want to conclude with this pun on Bruno Latour's famous title, however tempting a *bon mot*. In *We Have Never Been Modern* (1993 [1991]), Latour suggests that the North Atlantic's "modern constitution" rests upon a divide between scientific power, meant to represent things as they are, and political power, meant to represent subjects as they wish to be. Latour sees the formulation of this divide (science/politics, object/subject, nature/culture) as the impossible dream of modernity, since the world so neatly divided is actually made of hybrids. Nevertheless, Latour does admit, almost in passing, that blind faith in this divide also makes the moderns invincible. I am interested in this invincibility. Latour's witty title could be misread as to imply that we could have been modern according to definition. But if modernity is as much blind faith in this narrative as its global consequences, we have long been modern, except that the "we" here is not only the North Atlantic but also the hidden faces of a modernity necessary to North Atlantic hegemony — if not invincibility.

Ultimately, however, the fact that modernity has long obtained outside of the North Atlantic is only a secondary lesson from the Caribbean; it is a conclusion that still makes those outside of the North Atlantic the ones who need to be explained. Yet is the alter-native really what is to be explained? Is the puzzle the female slave who used her kerchief as individual signature, or the laws that repeatedly tried to curb her individual expression? Is the puzzle the resilience of the creolization process under slavery, or the expectation that enslaved Africans and their descendants would be either a *tabula rasa* or mere carriers of tradition (Trouillot 1998)? In short, is not the puzzle within the West itself?

The Caribbean story as I read it is less an invitation to search for modernity in various times and places—a useful yet secondary enterprise—than an exhortation to change the terms of the debate. What needs to be analyzed further, better, and differently is the relation between the geography of management and the geography of imagination that together underpinned the development of world capitalism and the legitimacy of the West as the universal unmarked. Anthropologists need to take further distance from North Atlantic universals as carriers of that legitimacy. As a discipline, we have launched the most sustained critique of the specific proposals rooted in these universals within academe. Yet we have not explored enough how much these universals set the terms of the debate and restricted the range of possible responses. In the context of this much-needed reformulation, the Caribbean's most important lesson is a formidable one, indeed. That lesson, as I see it, is that modernity never was— never could be—what it claims to be.

CHAPTER FOUR

Myths of Caribbean Identity

STUART HALL

Stuart Hall was born in 1932 in a middle-class family in Jamaica. He described his father as belonging to the 'coloured lower-middle-class... ethnically very mixed—African, East Indian, Portuguese, Jewish', and his mother's family as 'much fairer in colour' and 'English-oriented' (Hall 1996, 484). He went to the UK in 1951 where he studied at Oxford University, 'was saturated in West Indian expatriate politics' (Ibid., 492), met African students, and associated with what became known as the New Left-Alan Hall, Raymond Williams, Raphael Samuel, Peter Sedgewick and Perry Anderson. He became an editor of the Universities and Left Review which became the New Left Review. He moved to London in 1957 where he taught in a secondary school and edited the review until 1961. He taught media, film and popular culture at Chelsea College, University of London. He moved to Birmingham University in 1964 where he worked with, and then succeeded, Richard Hoggart as the director of the new Centre for Cultural Studies. In 1979, he became a professor of sociology at the Open University, an institution that pioneered the use of television and distance learning in higher education.

Hall's work has been important in the formation and development of cultural studies as an interdisciplinary and international field. He is well-known for his work on questions concerning the relations between race, ethnicity, culture and identity, particularly in connection with West Indians at home and abroad.

MYTHS OF CARIBBEAN IDENTITY

I want to talk about questions of culture and identity, specifically questions of Caribbean culture and identity, and to suggest that such questions are not in any sense separate or removed from the questions of political mobilisation, of cultural development, of national identification and of economic development. The more we know and see of the struggles of the societies of the periphery to make something of the slender resources available to them, the more important we understand the questions and problems of cultural identity to be in that process. I want to examine some of the themes of this enormous topic, which has been richly explored especially by Caribbean writers and artists-the question of cultural identity as it has presented itself always as a problem to Caribbean people.

Why it should be a problem is not a mystery, but I want to probe this question of identity and why Caribbean writers, politicians, civic leaders, artists and others have been unable to leave worrying this problem. And in doing so, I want to analyze the way we think about identity, and to explore the term 'myth' itself. The English are not good at myth; they always oppose it on the one hand to reality, on the other hand to truth, as if you nave to choose between them. I specifically do not want to choose between myth and reality, but to talk about the very real contemporary and historical effects of the myths of identity, and to do so with one other purpose which I hope will come through more clearly at the end. My own view is that the issue of cultural identity as a political quest now constitutes one of the more serious global problems as we go into the twenty-first century. The re-emergence of questions of cultural identity, of ethnicity, of nationalism the obduracy, the dangers and the pleasures of the rediscovery of identity in the modern world, inside and outside of Europe places the question of cultural identity at the very centre of the contemporary political agenda. What I want to suggest is that, despite the dilemmas and vicissitudes of identity through which Caribbean people have passed and continue to pass, we have a tiny but important message for the world about how to negotiate identity.

There is a very clear and powerful discourse about cultural identity, especially in the West... But the discourse of identity suggests that the culture of a people is at root (and the question of roots is very much at issue) a question of its essence, a question of the fundamentals of a culture. Histories come and go, peoples come and go, situations change, but somewhere down there is throbbing the culture to which we all belong. It provides a kind of ground for our identities, something stabilized, around which we can organise our identities and our sense of belonging. And there is a sense that modern nations and peoples cannot survive for long and succeed without the capacity to touch ground, as it were, in the name of their cultural identities.

The question of what a Caribbean cultural identity might be has been of extraordinary importance, both before but especially during the twentieth century, partly because of the dislocations of conquest, of colonisation and slavery, and partly because of the colonial relationship itself and the distortions of living in a world culturally dependent and dominated from some centre outside the place where the majority of people lived. But it has also been important for counter-identities, providing sources from which the important movements of decolonisation, of independence, of nationalist consciousness in the region have been founded. In a sense, until it is possible to state who the subjects of independence movements are likely to be, and in whose name cultural decolonisation is being conducted, it is not possible to complete the process. And that process involves the question of defining who the people are. In *Black Skin White Masks*, Fanon[1] speaks of what he calls 'a passionate research directed to the secret hope of discovering beyond self-contempt, resignation and abjuration, some beautiful and splendid area whose existence rehabilitates us both in regard to ourselves and others'. And as I have said, that passionate research by Caribbean writers, artists and political leaders, that quest for identity, has been the very form in which much of our artistic endeavours in all the Caribbean languages has been conducted in this century.

Why then is the identity of the Caribbean so problematic? It is a very large question, but let me suggest some of the reasons. First of all, if the search for identity always involves a search for origins, it is impossible to locate in the Caribbean an origin for its peoples. The indigenous peoples of the area very largely no longer exist, and they ceased to exist very soon after the European encounter. This is indeed the first trauma of identity in the Caribbean. How many of you know what the coat of arms of Jamaica is? It has two Arawak Indian figures supporting a shield in the middle, which is crossed by pineapples surmounted by an alligator. In 1983 the then Prime Minister of Jamaica, Edward Seaga, wanted to change the coat of arms on the ground that he could not find represented in it a single recognisable Jamaican identity. 'Can the crushed and extinct Arawaks,' he asked, 'represent the dauntless inhabitants of Jamaica? Does the low-slung near-extinct crocodile, a cold-blooded reptile, symbolise the soaring spirits of Jamaicans? Where does the pineapple, which was exported to Hawaii, appear prominently either in our history or in our folklore?' I read that quote simply to remind you that questions of identity are always questions about representation. They are always questions about the invention, not simply the discovery of tradition. They are always exercises in selective memory and they almost always involve the silencing of something in order to allow something else to speak...

Silencing as well as remembering, identity is always a question of producing in the future an account of the past, that is to say it is always about narrative, the stories which cultures tell themselves about who they are and where they came from. The one way in which it is impossible to resolve the problem of identity in the Caribbean is to try looking at it, as if a good look will tell you who the people are. During the period in which I was preparing my BBC series,[2] I had the occasion in a relatively short space of time to visit a large number of Caribbean islands, several of which I had not seen before. I was absolutely staggered by the ethnic and cultural diversity I encountered. Not a single Caribbean island looks like any other in terms of its ethnic composition, simply from the point of view of the different genetic and physical features and characteristics of the people. And that is before you start to touch the question of different languages and cultural traditions, which reflect diverse colonising structures.

It may be a surprise to some people that there are several Caribbean islands, large ones, in which blacks are nowhere near a majority of the population. There are now two important ex-British Caribbean nations where Asians are in a majority, and in Cuba, what strikes you first of all is the long persistence of white Hispanic settlement and then of the mulatto population, and only later of the black population. Haiti, which is in some ways the symbolic island of black culture, and where one feels closer to the African inheritance than anyone else, has a history in which mulattos have played an absolutely vital and key historical role. Martinique is a bewildering place, it is in my experience more French than Paris, just slightly darker. In the Dominican Republic it is possible to feel closer to Spain and to the Spanish tradition of Latin America than anywhere else I have been in the Caribbean. The melting-pot of the British islands produced everywhere you look a different combination of genetic features and factors, and in each island elements of other ethnic cultures Chinese, Syrian, Lebanese, Portuguese, Jewish are present. I know because I have a small proportion of practically all of them in my own inheritance. My inheritance is African, also I'm told Scottish of pretty low descent, probably convict East Indian, Portuguese, Jew. I cannot summon up any more but if I searched hard I expect I could find them.

What is more, in another sense, everybody there comes from somewhere else, and it is not clear what has drawn them to it, certainly not whether the motives were ever of the highest level of aspiration. That is to say, their true cultures, the places they really come from, the traditions that really formed them, are somewhere else. The Caribbean is the first, the original and the purest diaspora. These days blacks who have completed the triangular journey back to Britain sometimes speak of the emerging black British diaspora, but I have to tell them that they and I are twice diasporised. Furthermore, we are not just living in a diaspora where the centre is always somewhere else, but we are the break with those originating cultural sources as passed through the traumas of violent rupture. I don't want to speak about the trauma of transportation, of the breaking up of linguistic and tribal and familial groups; I don't want to talk about the brutal aftermath of Asian indenture. I simply want to say that in the histories of the migration, (forced or free, of peoples who now compose the populations of these societies, whose cultural traces are everywhere intermingled with one another), is the stamp of historical violence and rupture.

Of course, the peoples thus inserted into these old colonising plantation societies instantly polarised. And if anyone is still under the illusion that questions of culture can ever be discussed free and outside of questions of power, you have only to look at the Caribbean to understand how for centuries every cultural characteristic and trait had its class, colour and racial reference. You could read off from the populations to the cultures, and from the cultures to the populations, and each was ranked in an order of cultural power. It is impossible to approach Caribbean culture without understanding the way it was continually inscribed by questions of power. Of course that inscription of culture in power relations did not remain polarised in Caribbean society, but I now understand that one of the things I was myself running away from when I came to England to study in 1951 was a society that was profoundly culturally graded; that is what the old post-colonial society I grew up in was like. Of course those cultural relations did not remain fixed, and the relative cultures were quickly open to integration, assimilation and cross-influence. They were almost never self-contained. They became subject at once to complex processes of assimilation, adoption, adaptation, resistance, reselection and so on. That is to say, they became in a deep sense diasporic societies. For wherever one finds diasporas, one always finds precisely those complicated processes of negotiation and cross-influence which characterise Caribbean culture. I don't want to try and sketch the cultural relations of that period, but simply to identify three key processes which are at work creating the enormously refined and delicate tracery, the complexes of cultural identification, in Caribbean society in that time.

First, and especially with respect to the populations that had been enslaved, there has been the retention of old customs, the retention of cultural traits from Africa. Customs and traditions which were retained in and through slavery, in plantation, in religion, partly in language, in folk customs, in music, in dance, in all those forms of expressive culture which allowed men and women to survive the trauma of slavery. These were not intact, never pure, never untouched by the culture of Victorian and pre-Victorian English society, never outside Christianity or entirely outside the reach of the church, never without at least some small instruction in the Bible, always surrounded by the colonising culture, but importantly (and to some extent today imperatively) retaining something of the connection, often unrecognised, often only in practice, often unreflected, often not knowing that people were practising within a tradition. Nevertheless, in everyday life, insofar as it was possible, maintaining some kind of subterranean link with what was often called the other Caribbean, the Caribbean that was not recognised, that could not speak, that had no official records, no official account of its own transportation, no official historians, but nevertheless had an oral life which maintained an umbilical connection with the homeland and the home culture.

But let us not forget that retention characterised the colonising cultures as well as the colonised. For if you look at the Little Englands, the Little Spains and the Little Frances that were created by the colonisers, if you consider this kind of fossilised replica, with the usual colonial cultural lag—people are always more-Victorian when they're taking tea in the Himalayas than when they're taking tea in Leamington—they are keeping alive the memory of their own homes and homelands and traditions and customs. This very important double aspect of retention has marked Caribbean culture.

Secondly, the profound process of assimilation, of dragging the whole society into some imitative relationship with this other culture which one could never quite reach. When one talks about assimilation in the Caribbean, one always feels Caribbean people constantly leaning forward, almost about to tip over, always just going somewhere else. My mother used to tell me that if she could only get hold of the right records, she would be able to stitch together a kind of genealogy for her household-not one that led to the West Coast of Africa, believe me, but a genealogy which would connect her, she wasn't quite sure, to the ruling house of the Austro-Hungarian empire or the lairds of Scotland, one way or the other. She probably thought that maybe in the quadrangle of Merton College, Oxford I might stumble across one of these secret stones that would somehow convert me into what clearly I was formed, brought up, reared, taught, educated, nursed and nurtured to be, a kind of black Englishman. When I first went home in the mid 1950s, my parents said to me, 'I hope they don't take you to be one of those immigrants over there.' And the funny thing is, I'd never called myself, or thought of myself as an immigrant before. But having once been hailed or interpellated, I owned up at once; that is what I am. In that moment I migrated. Again, the word black had never been uttered in my household or anywhere in Jamaica in my hearing, in my entire youth and adolescence-though there were all kinds of other ways of naming, and large numbers of people were very black indeed-so it was not until the mid 1960s, on another visit home, that my parents said to me, 'There's all this black consciousness, black movement in the United States, I hope it's not having an influence over there', and I realised I had just changed identity again. I owned up once more and said, 'Actually you know, I am exactly what in Britain we are starting to call black'. Which is a sort of footnote to say, identity is not only a story, a narrative which we tell ourselves about ourselves, it is stories which change with historical circumstances. And identity is far from the way in which we think and hear them and experience them. Far from only coming from the still small point of truth inside us, identities actually come from outside, they are the way in which we are recognised and then come to step into the place of the recognitions which others give us. Without the others there is no self, there is no self-recognition.

So given the skewed structures of growing up in such a society, of attempting whatever social rank or position in the racial colour structure you occupy, of trying to negotiate the complexities of who out of these complicated sets of stories you could possibly be, where you could find in the mirror of history a point of identification or recognition for yourself, it is not surprising that Caribbean people of all kinds, of all classes and positions, experience the question of positioning themselves in a cultural identity as an enigma, as a problem, as an open question. There are many writings about this question, but for me the overwhelmingly powerful statement is in Fanon's book *Black Skin White Masks*, for only in Fanon do you understand the internal traumas of identity which are the consequence of colonisation and enslavement. That is to say, not just the external processes and pressures of exploitation, but the way that internally one comes to collude with an objectification of oneself which is a profound misrecognition of one's own identity. Consequently, against that background, in the New World and in the Caribbean, the attempts in the twentieth century to reach for independence, to decolonise, the movements in the nineteenth century in the Hispanic Caribbean societies for independence from Spain, the attempts to regenerate and ground the political and social life of the society, not in an absent picture or image which could never be fulfilled, not in the nostalgia for something outside the society, but in the complicated realities and negotiations of that society, itself, is a question which had to entail the redefinition of identity. Without it there could have been no independence of any kind. And one of the complexities or perplexities of the independence movement certainly in the British Caribbean islands is that in my view in the early phases of those movements so-called political independence from the colonial power occurred, but the cultural revolution of identity did not.

For the third process, which will form the rest of my talk, I want to start by looking at some of the other attempts to name the unnameable, to speak about the possibilities of cultural identification, of the different traditions of the peoples for whom on the whole there were no cultural models, the peoples at the bottom of the society. And as you can imagine, that always involved a renegotiation, a rediscovery of Africa. The political movements in the New World in the twentieth century have had to pass through the re-encounter with Africa. The African diasporas of the New World have been in one way or another incapable of finding a place in modern history without the symbolic return to Africa. It has taken many forms, it has been embodied in many movements both intellectual and popular. I want to say a word about two or three of them only. Perhaps best known in an intellectual sense is the movement around the notion of *négritude*, around the discovery of blackness, the affirmation of an African personality, very much associated with the name of Aimé Césaire,[3] and of the group around Césaire in Paris and afterwards, coming out of Martinique, a tiny society which I described earlier on in a rather pejorative way, the most French place I have encountered in the Caribbean, certainly, but also the home-place of both Fanon and of Aimé Césaire. Césaire's work lay in plucking out of that Caribbean culture with which he was most familiar the strands that related most profoundly back to the valorisation of the African connection, the rediscovery of the African connection, of African consciousness, of African personality, of African cultural traditions.

I was fortunate enough in the programme on Martinique to be able to include an interview with Aimé Césaire, who must be nearly twice my age and looks about half of it, wonderfully fit and resilient at this moment. In that interview you can see the enormous pleasure with which he describes the story of having gone to Africa and rediscovered for the first time the source of the masks of the Martinique carnival which he had played in and helped to make when he was a boy-suddenly the flash of recognition, the continuity of the broken and ruptured tradition. The enormously important work that flowed from his involvement in the *négritude* movement, not only the poems and the poetry and the writing which has come out of that inspiration, of the renegotiation of a Caribbean consciousness with the African past, but also the work which he has inspired in Martinique amongst poets and painters and sculptors, is a profound revelation of how creative this symbolic reconnection has been. And yet of course the paradox is that when Aimé Césaire opens his mouth you hear the most exquisitely formed lycée French. I hardly know anyone who speaks a more perfect French, it is beautifully articulated. 'I am', he says, 'French, my mind is French'. Looking out for the right word, he says, 'like if you went to Oxford you would be English. I went to a French school, I was taught the French language, I wasn't allowed to use kréyole at home, I learned only French classical culture, there's a strong tradition of assimilation, I went of course to Paris where all bright young Martiniquans went'. And because of the tradition of political assimilation, he has in fact done what no black British Caribbean has ever done, which is to sit in the parliament of his own metropolitan society. Nevertheless when Aimé Césaire started to write poetry, he wanted, because of his interest, alerted and alive to the subterranean sources of identity and cultural creativity in his own being, to break with the models of French classical poetry. And if you know his notebook, the *Return to My Native Land,* you will know how much that is a language which in its open roaring brilliance, has broken free from those classical models. He becomes a Surrealist poet Aime' Césaire has never, as you perhaps know, argued for the independence of Martinique. Martinique has a very particular position, it is an internal department of France, and those of you who want to be crude and materialist about it had better go and see the kinds of facilities which that gives Martiniquan people, and compare them with the facilities available to most of the other peoples of the Caribbean islands, before you begin to say what a terrible thing this is. Nevertheless, my own feeling, though I have no enormous evidence for this, is that the reluctance of Césaire to break the French connection is not only a material one but a spiritual one. He went to the Schoelcher lycée. Schoelcher[4] was an important early Martiniquan figure, and in celebrating an anniversary of Schoelcher, Césaire said, 'He associated in our minds the word France and the word liberty, and that bound us to France by every fibre of our hearts and every power of our minds'. He said, 'I know only one France, the France of the Revolution, the France of Toussaint L'Ouverture. So much for gothic cathedrals'.

Well, so much-indeed for gothic cathedrals. The France with which Césaire identifies, and it has played of course a most profound role in Caribbean history, is one France and not another, the France of the Revolution, the France *of liberté, egatité, fraternité,* the France that

Tousssaint L'Ouverture heard, the France that mobilised and touched the imagination of slaves and others in Haiti before the Revolution. And yet in the actual accounts of the revolution that we have, one of the most difficult, one of the trickiest historical passages to negotiate is precisely how much in the spark of the various things that went into the making of the Haitian Revolution can be attributed on the one hand to the ruptures sweeping out in the wake of the French Revolution, and on the other hand to the long experience of a severe and brutal regime on the plantations themselves, what you might call the revolutionary school of life itself. There were also, of course, the traditions of Africa and of African resistance, and of *marronage* in the plantation villages themselves. We don't know. It is an impossible enigma to sort out, in one of the momentous historical events of Caribbean history, to what can be attributed the different elements that come together in that revolutionary conjuncture.

Césaire was influenced in part by his contact at an early stage with an important movement in the United States which now goes under the title of the Harlem Renaissance.[5] I don't know how much you know about the writers of the Harlem Renaissance, of Langston Hughes and Countee Cullen and van Vechten, an important movement of writers, intellectuals and artists in New York in the early phases of the twentieth century, that had an important influence on a variety of Caribbean writers, poets and artists. One of the important things that the movement of the Harlem Renaissance did, was on the one hand to speak about the importance and the distinctiveness, the cultural and aesthetic distinctiveness of the black American contribution to American culture. The other important thing that movement did was to stake a claim for American blacks in the centre and at the heart of modernism itself. The writers of the Harlem Renaissance did not wish to be located and ghettoised as ethnic artists only able to speak on behalf of a marginal experience confined and immured in the past, locked out of the claim to modern life. What they said was, the experience of blacks in the New World, their historical trajectory into and through the complex histories of colonisation, conquest and enslavement, is distinct and unique and it empowers people to speak in a distinctive voice. But it is not a voice outside of and excluded from the production of modernity in the twentieth century. It is another kind of modernity. It is a vernacular modernity, it is the modernity of the blues, the modernity of gospel music, it is the modernity of hybrid black music in its enormous variety throughout the New World. The sound of marginal peoples staking a claim to the New World. I say that as a kind of metaphor, just in case you misunderstood the point I was trying to make about Aimé Césaire. I am anxious that you don't suppose I see him as an assimilationist Frenchman, deeply in bad faith because he is invoking Africa. I am trying to do something else. I am talking about the only way in which Africa can be relived and discovered by New World blacks who are diasporised irrevocably, who cannot go back through the eye of the needle.

Let me talk about finally going back through the eye of the needle. There was a very famous moment during the explosion of Rastafarianism in Jamaica in the sixties when a somewhat beleaguered Prime Minister said, 'Well perhaps you ought to go back to Africa. You've talked about it so much, you say you came from there, you say you're still in slavery here, you're not in a free land, the promised land is back there where somebody took you from, perhaps you ought to go back and see'. Well, of course some people did go back and see, as you perhaps know. Of course they did not go back to where they came from, that was not the Africa they were talking about. Between the Africa that they came from and the Africa that they wanted to go back to, two absolutely critical things had intervened. One is, that Africa had moved on. Africa—one has to say it now and again to somewhat nostalgic and sentimental nationalists in the Caribbean-Africa is not waiting there in the fifteenth or seventeenth century, waiting for you to roll back across the Atlantic and rediscover it in its tribal purity, waiting there in its prelogical mentality, waiting to be awoken from inside by the returning sons and daughters. It is grappling with the problem of AIDS and underdevelopment and mounting debt. It is trying to feed its people, it is trying to understand what democracy means against the background of a colonial regime which ruptured and broke and recut and reorganised peoples and tribes and societies in a horrendous shake-up of their entire cognitive and social world. That is what it's trying to do, twentieth-century Africa. There is no fifteenth-century mother waiting there to succour her children. So in that literal sense, they wanted to go somewhere else, they wanted to go to the other place that had intervened, that other Africa which was constructed in the language and the rituals of Rastafarianism.

Now as you know, the language and rituals of Rastafarianism speak indeed of Africa, of Ethiopia, of Babylon, of the Promised Land, and of those who are still in suffering. But like every chiliastic language which has been snatched by the black people of the New World diasporas out of the jaws of Christianity, and then turned on its head, or read against the grain, or crossed by something else—and the New World is absolutely replete with them—it is impossible in my experience to understand black culture and black civilisation in the New World without understanding the cultural role of religion, through the distorted languages of the one book that anybody would teach them to read. What they felt was, I have no voice, I have no history, I have come from a place to which I cannot go back and which I have never seen. I used to speak a language which I can no longer speak, I had ancestors whom I cannot find, they worshipped gods whose names I do not know. Against this sense of profound rupture, the metaphors of a new kind of imposed religion can be reworked, can become a language in which a certain kind of history is retold, in which aspirations of liberation and freedom can be for the first time expressed, in which what I would call the imagined community of Africans can be symbolically reconstructed.

I said to you that when I left Jamaica in the 1950s it was a society which did not and could not have acknowledged itself to be largely black. When I went back to Jamaica at the end of the sixties and in the early seventies, it was a society even poorer than when I had left it, in material terms, but it had passed through the most profound cultural revolution. It had grounded itself where it existed. It was not any longer trying to be something else, trying to match up to some other image, trying to become something which it could not. It had all the problems in the world sticking together, finding the wherewithal to get to the next week, but in terms of trying to understand ordinary people—I'm not now talking about intellectuals, I'm talking about ordinary people—the important thing was the new realisation that they could speak the language that they ordinarily spoke to one another anywhere. You know, the biggest shock for me was listening to Jamaican radio. I couldn't believe my ears that anybody would be quite so bold as to speak patois, to read the news in that accent. My entire education, my mother's whole career, had been specifically designed to prevent anybody at all, and me in particular, from reading anything of importance in that language. Of course, you would say all kinds of other things, in the small interchange of everyday life, but important things had to be said, goodness knows, in another tongue. To encounter people who can speak with one another in exactly that transformation of standard English which is patois, which is Creole—the hundreds of different Creole and semi-creole languages which cover the face of the Caribbean in one place or another—that these have become as it were the languages in which important things can be said, in which important aspirations and hopes can be formulated, in which an important grasp of the histories that have made these places can be written down, in which artists are willing for the first time, the first generation, to practise and so on, that is what I call a cultural revolution.

And it was in my view made by the cultural revolution of Rastafarianism. What I mean by that is certainly not that everybody became Rasta, although there was a moment in the sixties there when it was pretty hard not to be. I once interviewed a very old Rastafarian figure about the large numbers of Kingston intellectuals and students who were growing their locks down to their ankles. And I said to him, as part of a long interview about the nature of Rastafarianism, how he'd got into it, and so on, 'What do you think of these weekend Rastas, these middle-class Rastas, do you think they're up to anything, do you think they can reason?' and he said, 'You know, I don't say anything about them, I don't think anything about them, because in my church everybody reasons for themselves. So if they want to reason in that way, that's their business'. Well I thought that was a nice gentle remark, but I wanted to nail him, so I said, 'Listen to me now, isn't Haile Selasse dead, so the bottom has just fallen out of this whole Rastafarian business? He's dead, how can the Son of God be dead?' And he said to me, 'When last you hear the truth about the Son of God from the mass media?'

You see, it was not the literal place that people wanted to return to, it was the language, the symbolic language for describing what suffering was like, it was a metaphor for where they were, as the metaphors of Moses and the metaphors of the train to the North, and the metaphors of freedom, and the metaphors of passing across to the Promised Land, have always been metaphors, a language with a double register, a literal and symbolic register. And the point was not that some people, a few, could only live with themselves and discover their identities by literally going back to Africa—though some did, not often with great success—but that a whole people symbolically re-engaged with an experience which enabled them to find a language in which they could describe and appropriate their own histories.

I want to close. I have said something about the intellectual movement of *négritude,* I've referred to another important movement, not in the Caribbean but with influence on the Caribbean, the Harlem Renaissance of the twenties, and I've talked about the cultural revolution in the wake of Rastafarianism. One of the most important things that people on this side of the Atlantic know about Rastafarianism is that it produced the greatest reggae artist in the world, Bob Marley.[6] And I think many Europeans believe that reggae is a secret African music that we've had tucked in our slave knapsacks for three or four centuries, that we hid out in the bush, practised at night when nobody was looking, and gradually as things changed we brought it out and began to play it a little, feed it slowly across the airwaves. But as anybody from the Caribbean would know, reggae was born in the 1960s. Actually it was the answer to ska. When I returned to Jamaica I heard these two musical traditions. In *The Invention of Tradition,* the collection edited by Eric Hobsbawm and Terence Ranger,[7] it's explained that many British traditions which people believed have been around since Edward I were actually developed by Elgar or Disraeli,[8] the day before yesterday. Well, reggae is a product of the invention of tradition. It is a sixties music, its impact on the rest of the world comes not just through preservation—though it is rooted in the long retained traditions of African drumming—but by being the fusion, the crossing, of that retained tradition with a number of other musics, and the most powerful instruments or agencies of its world propagation were those deeply tribal instruments, the transistor set, the recording studio, the gigantic sound system. That is how this deeply profound spiritual music of Africa that we've been treasuring got here.

It's not part of my story to tell what it did here in Britain, but actually it not only provided a kind of black consciousness and identification for people in Jamaica, but it saved the second generation of young black people in this society. Is this an old identity or a new one? Is it an ancient culture preserved, treasured, to which it is possible to go back? Is it something produced out of nowhere? It is, of course, none of those things. No cultural identity is produced out of thin air. It is produced out of those historical experiences, those cultural traditions, those lost and marginal languages, those marginalised experiences, those peoples and histories which remain unwritten. Those are the specific roots of identity. On the other hand, identity itself is not the rediscovery of them, but what they as cultural resources allow a people to produce. Identity is not in the past to be found, but in the future to be constructed.

And I say that not because I think therefore that Caribbean people can ever give up the symbolic activity of trying to know more about the past from which they come, for only in that way can they discover and rediscover the resources through which identity can be constructed. But I remain profoundly convinced that their identities for the twenty-first century do not lie in taking old identities literally, but in using the enormously rich and complex cultural heritages to which history has made them heir, as the different musics out of which a Caribbean sound might one day be produced...

NOTES

1. See chapter 17 in this volume.

2. Hall produced a seven-part television series about Caribbean culture called 'Redemption Song' for the British Broadcasting Corporation (1991).

3. See chapter 16 in this volume.

4. Victor Schoelcher (1804-93), the son of a French porcelain manufacturer, encountered slavery first-hand when his father sent him to the Caribbean in 1829 in search of colonial outlets. He wrote accounts of what he saw to a Paris magazine, describing slave markets and punishments, and advocating the gradual abolition of slavery. He became the leading French abolitionist and, when he became a member of the revolutionary government of 1848, he organised emancipation. The freed slaves were declared citizens of France and adult males had voting rights in the Second Republic (1848-51). Schoelcher fought against the coup of Louis Napoleon, which ended their rights along with the republic, and he was exiled from France in 1852.

5. The African-American cultural movement centred in New York City and known as the Harlem Renaissance was influenced by Marcus

Garvey's ideas. One of its major figures was the Jamaican poet and novelist Claude McKay (1890-1948), author of *Banjo* (1929), *Banana Bottom* (1933) and many other works.

6. Bob Marley (1945-81), a Jamaican singer and songwriter, was the first reggae performer to achieve international stardom. He joined a group, the Rudeboys, in 1961, that later became the Wallers. By 1967, when Marley became a Rastafarian, reggae had emerged from a succession of musical styles, including ska, which incorporated elements of rhythm and blues from the southern United States and a kind of Jamaican folk song called mento. Many of Marley's songs convey messages of cultural revolution and spiritual redemption. His albums include 'Catch a Fire' (1972), 'Natty Dread' (1975), 'Rastaman Vibrations' (1976), 'Exodus' (1977), 'Kaya' (1978) and 'Uprising' (1979).

7. Eric Hobsbawm and Terence Ranger, editors, *The Invention of Tradition* (New York: 1983).

8. Sir Edward Elgar (1857—1934) was a distinguished English romantic composer whose popular 'Pomp and Circumstance' marches came to symbolise the British Empire. Benjamin Disraeli (1804—81), a British writer and prime minister, shaped the modem conservative party and supported traditional institutions such as the monarchy. He emphasised Britain's imperial policy by creating the title 'Empress of India' for Queen Victoria in 1876.

CHAPTER FIVE

The Golden Age of Piracy

JENNIFER MARX

*Roberts himself made a gallant figure at the time of the engage-
ment, being dressed in a rich crimson damask waistcoat and
breeches, a red feather in his hat, a gold chain around his neck,
with a diamond cross hanging to it, a sword in his hand and
two pair of pistols, hanging at the end of a silk sling, flung over
his shoulders.*

Captain Charles Johnson, 1724.

Several chapters stand out in the bloody annals of seaborne
villainy, but none boasts more fascinating exploits than the brief
but brilliant period around the beginning of the eighteenth century
known as the Golden Age of Piracy. For thirty years or so, pirates
in record numbers sallied forth from nerve centers in the Caribbean
and North American ports to prey on the burgeoning mercantile
traffic sailing between Europe and America and traffic from Africa
and the East.

The great flare-up of maritime lawlessness that convulsed the
Caribbean and the North American coast involved pirates of various
nationalities. The Virgin Islands harbored outlaws from many
countries and New Providence in the Bahamas was the nest of a
swarm of British and American pirates. Spanish pirates operated out
of Cuba, Puerto Rico, and St. Augustine. French pirates continued
to use Martinique as their main base.

The Golden Age furnished the classic pirate images immortalized by painters, writers, and film makers. Most of the pirates we are familiar with today were members of this international fraternity who prowled the seas until the Admiralty trials in 1722, when the hanging of fifty-two of Captain Roberts's men at Cape Coast Castle on the African coast brought the curtain down on one of history's greatest outbreaks of sea robbery. Pirates continued to plunder shipping for another century but without the devastating effects of previous eras and certainly without providing such flamboyant characters.

Figure 5.1: Blackbeard in Smoke and Flame, *a painting by Frank Schoonover commissioned to illustrate an article in* American Boy *magazine in September 1922. All the descriptions of Blackbeard suggest that he was as terrifying as he appears here, and that he deliberately cultivated a fearsome image in order to keep his crew in their place and to encourage his victims to surrender without fight.*

The pirates were a motley lot: men of many nations, several races, varied abilities, and widely ranging temperaments. The following description fits the typical pirate of the age: "his parentage was but meane, his estate lowe, and his hope less." Some, however, like Major Stede Bonnet, the runaway planter, were well born. Others were well educated: Dr. John Hincher, a graduate of the University of Edinburgh, was tried for piracy in Newport, Rhode Island, in 1723. He was acquitted when he testified that he had been forced against his will to act as ship's surgeon by the infamous Captain Low.

Pirates were the darlings of the eighteenth-century media and the public avidly followed their exploits. Their many feats and foibles, recorded in contemporary chronicles and trial proceedings, were embellished upon in broadsides and handbills that the public lapped up. A pirate trial was always mobbed with spectators, but what they loved best was a good hanging. Many of the rogues went to the gallows displaying the dash that had marked their careers. Finely dressed in silk and velvet, a condemned man might sprinkle the mob with gold coins or pearls before he swung off.

The spectators waited with bated breath to hear the "last words" the condemned addressed to the crowd before mounting the gallows. They applauded the pirates who "died well," jeered those who didn't, and generally enjoyed themselves.

WAR BRINGS PRIVATEERS

England and Spain were inevitably drawn into the wars of the eighteenth century, which were primarily concerned with maintaining the balance of power in Europe. Just as inevitably, these conflicts spread to the Spanish Indies. At intervals throughout the century, Anglo-French controversy over control of Spain's New World dominions and over the monopoly of trade erupted in naval warfare, which was played out in the Caribbean and the western Atlantic. During wars pirates enjoyed the sanction of privateering licenses. As soon as peace treaties were signed far across the Atlantic, "the West Indies always swarm with pirates," a colonial official in the Bahamas complained in 1706.

As early as 1695 the governor of Jamaica had written the king of England that so many men were involved in piracy that it was impossible to find crews for merchant ships anywhere in the West Indies. A few years later, following the Treaty of Ryswick, which ended war between England and France, thousands of seamen and privateers were out of work and eager to go "on the account."

PEACE BRINGS PIRATES

Times were hard after the Treaty of Utrecht ended the War of Spanish

Figure 5.2: A *double-headed shot for cutting through the rigging of an enemy ship. This particular example was fired during the Battle of Trafalgar and killed eight men onboard Admiral Nelson's* Victory.

Succession in 1713, bringing a quarter century of peace to Europe but unleashing a horde of hungry seamen in tarred breeches and canvas jackets who couldn't find honest employment. During the war they had pillaged and plundered with official sanction. After the war many turned pirate, continuing hostilities "beyond the line" as both France and England vied to fill the vacuum left by the erosion of Spanish domination, turning a blind eye to the pirates as long as they furthered their geopolitical objectives. Hollander pirates were notably absent from the action because the Dutch government wisely paid their skilled troops to work in the herring fleet so they could be marshaled for the next inevitable conflict.

The surge in piracy was predictable. Veterans' prospects were far from rosy in the West Indies where slave labor had severely restricted opportunity for the average white man. Navies and mercantile shipping never had enough berths in peacetime to absorb great numbers of men. What places there were in peacetime offered low wages and harsh conditions.

The colonies in the West Indies and along the American seaboard prospered after the war. At the peace, England wrested

Figure 5.3: *A final fling of the officially tolerated French buccaneers—and perhaps the greatest—was their attack on the Spanish treasure port of Cartagena in April J 697. A fleet of warships led by Baron de Poinds joined forces with French buccaneers under the command of Jean Baptise du Casse and launched a massive attack on the heavily fortified city. While the ships bombarded the forts, the troops laid siege to them. On May 6 the governor surrendered after securing terms that would permit the citizens to keep their possessions. Even so, the French ships were able to sail away on May 31 with their holds filled with treasure. They did so hastily, fearing that Spanish and English forces were en route.*

from Spain the *asiento,* or monopoly for supplying slaves to the Spanish colonies, along with their other commercial privileges. Increasing volumes of English, French, and Dutch colonial trade provided attractive piratical opportunities. Spain resumed scheduled sailings of the treasure fleet, making the area a more rewarding "cruising ground" than it had been for over a century. Consequently, in 1717 James Logan, colonial secretary of Pennsylvania, estimated there were at least 1,500 pirates lying in wait off the North American coast. The same year the governor of Bermuda lamented that "North and South America are infested with the rogues."

AN INFESTATION OF ROGUES

Who were these "rogues"? Judging from records, including trial records, the majority were lower-class men with roots in England's West Country and Wales, where piracy was in the blood. Nearly half of the fifty-two members of Black Bart Roberts's crew hanged at Cape Coast Castle on Africa's Gold Coast in 1722 were Welsh or West Countrymen. Others were former indentured servants or poor colonial whites. Then there were the Catholics. National ambitions pitted Protestant countries against Catholic Europe, so Catholic Irish and Scots pirates often signed on with Spanish and French expeditions rather than with British crews. Protestant Frenchmen served aboard English pirate ships.

On many ships there were runaway slaves and mulattos. There is even a record of a Native American from Martha's Vineyard who sailed with the pirate Captain Charles Harris. Pirates might invite slaves to join them if there were only a few aboard a prize. However, in the case of a slave ship, the entire cargo was generally marketed to unscrupulous dealers. For twenty years a band of Scots, English, Spaniaids, Portuguese, blacks, and mulattos was headquartered in the Bahamas under the leadership of Spaniard Agostino Blanco. Other nationalities included Dutch, Scandinavians, Greeks, and East Indians. Most pirates were single and not particularly tied to any one place. Pirate crews often excluded married men to avoid potential domestic entanglements.

Golden Age pirate captains might be in their thirties or forties but the average pirate was about twenty-five, young by today's standards but not in the eighteenth century, when a man of thirty was regarded as relatively old. Sailors were generally able-bodied seamen by age seventeen. An arrest warrant duly issued in December 1699 by Governor Nicholson of Virginia, one of the few colonial authorities genuinely committed to suppressing piracy, gives a vivid glimpse of a pirate crew. Among the outlaws at large in Virginia, and wanted for plundering in the Red Sea on the *Adventure*, were: "Tee Wetherlly, short, very small, blind in one eye, about eighteen; Thomas Jameson, cooper, Scot, tall, meagre, sickly look, large black eyes, twenty; William Griffith, short, well set, broad face, darkest hair, about thirty." There was another thirty-year-old, John Loyd, "of ordinary stature, rawboned, very pale, dark hair, remarkably deformed in the lower eyelid," as well as two children, one aged fifteen and the other "Thomas Simpson, short and small, much squint-eyed, about ten of age."

A SEAMAN'S LOT

Almost all English-speaking pirates were recruited from the ranks of the British navy or had served as merchant seamen or privateers. In comparison, life as member of a pirate crew was tantamount to a holiday cruise. It is hard to imagine how truly horrible the conditions for ordinary seamen were aboard navy and merchant vessels. Many of the men had been unwillingly seized, or pressed, and at times when vessels were in port the men were shackled to keep them from abandoning ship. "So much disabled by sickness, death, and desertion of their seamen" were His Majesty's ships, noted an officer, that they could scarcely combat the pirates. According to some estimates, 50 percent of all British and American men pressed between 1600 and 1800 died at sea.

Figure 5.4: Extorting Tribute from the Citizens, *by Howard Pyle, painted to illustrate his article "The Fate of a Treasure-Town" published in* Harper's Monthly Magazine *in December 1905. Du Casse's buccaneers had felt cheated at the reward distributed to them by de Pointis. A renegade group of them returned and managed to extract further riches by torture and threats.*

A PIRATE'S PERKS

Piracy offered immediate escape. When a merchant vessel was seized, the pirates often gave seamen the option to join them. Most did so willingly. The pirates questioned the men about their officers, especially the captain and the quartermaster. Officers deemed unusually cruel or otherwise despicable were usually killed, frequently after torture. Otherwise they might be marooned, set afloat in a small boat, left in a burning ship, or even ransomed.

Piracy offered more than release from a floating hell. A rover's life wasn't easy, but at least as part of a pirate crew he was an equal among equals, a member of an organized egalitarian community in which decisions were made collectively. Pirates were risk-sharing

Figure 5.5: Action off Cartagena, May 28, 1708, *by Samuel Scott. Just a few years after the French sacking, a Bntish squadron attacked the Spanish treasure fleet in the vicinity of the town. One ship was captured and another was driven ashore, but the* San Jose, *carrying the bulk of the treasure, blew up and sank due to the detonation of her powder magazine.*

partners enjoying a kind of freedom unthinkable ashore or in lawful sea service. Before setting out on a voyage, or upon election of a new captain, crew members subscribed to a set of written articles that forged a strong esprit de corps. They were derived from privateering articles and sworn to over a Bible or boarding ax and governed life aboard ship. They defined responsibilities, specified disciplinary measures, and dictated terms of the share-out of plunder.

Pirates of this era also established an early welfare system similar to that pioneered by the buccaneers. A portion of plunder was set aside in a fund to compensate men who were disabled during an engagement.

ONE FOR ALL AND ALL FOR ONE

Pirates gloried in turning the rigid stratification and tyrannical organization of the navy and merchant service upside down. The pirate captain was elected by the crew and served at their pleasure. He was commander only "in fighting, chasing, being chased," and could be deposed by a vote for cruelty, cowardice, refusal to go after a potential prize, or any other reason. The men elected a quartermaster who was their representative and who distributed food and cash and acted as ombudsman, adjudicating any disputes between them. The quartermaster often became captain of any prize vessel deemed worth keeping.

Perhaps one of the things pirates most appreciated was the Pirate Council, which was the highest authority for each pirate voyage, deciding where to go, what to do, and how to deal with problems. It paralleled the traditional council of war on naval vessels, in which only the highest-ranking officers participated, but with one huge difference: every man was a member.

PIRATE ARTICLES

Articles varied only slightly from captain to captain. The articles sworn to by the pirates signing on with Bartholomew Roberts are typical and appear in Johnson's book. They reveal democracy, equality, and discipline.

Captain Johnson wrote that the original signed articles had been thrown overboard when Roberts was captured, hinting that there were clauses that contained "something too horrid to be disclosed to any except such as were willing to be sharers in the iniquity of them."

One of the articles on Captain George Lowther's ship stipulated that the man who first spied a sail should have the best pistol or small arm on her. Fire was an ever-present danger, often addressed in a set of articles. Article VI abpard Captain John Phillips's *Revenge* stated: "That man that shall snap his arms, or smoke tobacco in the hold without a cap to his pipe, or carry a candle lighted without a lanthorn, shall suffer the same punishment as in the former article." The punishment referred to was "Moses's Law" or thirty-nine stripes on a man's bare back.

The captain had few special privileges, receiving the same rations as his mates, and having to share his cabin with whoever walked in. Men ate where they wished and slept wherever they wanted. The captain, quartermaster, and skilled "sea artists" including surgeons, carpenters, and boatswains received a greater share of plunder than the common pirates. So did musicians, whose loud (and often cacophonous) renderings served to urge the outlaws on and unnerve the enemy.

PRISONERS

Pirates were vengeful and often vented feelings of rage at the injustice of life on captives of higher birth and particularly on officers who represented the arbitrary authority they had been subject to on men-o'-war and merchant ships. When Philip Lyne was caught in 1726, he boasted that he had killed thirty-seven "Masters of Vessels" during "the time of his Piracy."

Some pirates were particularly sadistic. One of the worst was Captain Edward Low. When the captain of a Portuguese merchantman dropped a large bag filled with gold *moidores* (coins) overboard, he had the man's lips cut off and broiled in front of him and then forced the Portuguese mate to eat them before slaughtering the entire crew. Although pirates kept many prize vessels, generally modifying them to fit their needs, they rifled unsuitable vessels and then often set them ablaze to drift, sometimes with captives tied to the

Figure 5.6: *An eighteenth-century gravestone from Skipness in Kintyre, Scotland. The skull and crossed bones were often used to symbolize death—an image adopted by pirates for their flags. The hourglass and other symbols represent the passing of time and the limited span of our lives on earth; it was also incorporated into pirate flags to warn that the time for surrendering was limited.*

mast. The psychopathic Captain Low once took everyone off of a captured French prize except the cook, whom he strapped to the mast before setting the vessel on fire, because "being a greasy fellow" he would fry well.

Women often received some protection; however, the cost might mean accepting the amorous advances of a particular pirate, and there are records of women suffering barbarous atrocities. Prisoners were tortured for a variety of reasons: chief among them, to make them reveal hidden valuables or because there was no valuable cargo. Prisoners were sometimes hauled up in the rigging by block and tackle and then dropped to their death on the deck. Naked captives might be used as target practice.

A rather nasty pirate game was called "sweating." It involved arranging a circle of lighted candles on deck and forcing one captive after another to run around and around between the candles and the mast while each pirate, armed with a knife, fork, or sword, ran "his Instrument into his Posteriors" until the prisoner collapsed; all to the accompaniment of horn, drum, and violin.

On occasion pirates showed clemency and even kindness to officers who were well spoken of by their men. But it wouldn't do to be too nice; after all, pirates relied on their ability to inspire terror to hasten the capitulation of a vessel. In addition to pirates' fearsome reputation for torture, they employed the icon of piracy, the grisly Jolly Roger; still *jolie* (to the pirates) but no longer *rouge* or red, as the buccaneers' had been. The mere glimpse of a pirate pennant on the mast of an approaching ship was notice that "no quarter will be given" and paralyzed many a vessel.

THE JOLLY ROGER

The banner of "King Death" was also a potent symbol of pirate solidarity. In an age when children could be hanged for stealing a loaf of bread, the specter of the noose was hardly a deterrent to men who scorned society and all its laws, and thumbed their noses at death.

Flags during the Golden Age varied but generally had in common a black ground with symbols representing death, violence, and time running out for pirates' prey. The skull and crossbones image appeared on seventeenth-century English tombstones, and at sea when a man died the ship's captain drew a skull next to the man's name in the vessel's log. A number of pirate standards featured a skeleton toasting death with a raised rum glass. The glass, called a rummer, was popular in English taverns during the period when elaborate rum drinks were popular, but pirates in the Caribbean drank almost exclusively from pewter mugs or coconut shell cups.

BEWARE AND TAKE CARE

The pennant of Captain Low, a blood-red skeleton on a black ground, was especially feared. Low and George Lowther, with whom he sailed early in his career, represent the acme of senseless pirate cruelty. Low was born in Westminster, England, and from his childhood was a bully and a thief. After a stint as a merchant seaman he worked briefly in a Boston shipyard before being fired for antisocial behavior. Like so many others who couldn't make a go of it elsewhere, Low found his way to the Bay of Honduras to cut logwood. During a violent argument with his boss, Low shot a bystander and fled the camp. He and a dozen companions seized a small vessel and set out to "make a black Flag and declare War against all the World."

Lowther began as a second mate on a Royal Africa Company ship carrying slaves. Sailors loathed voyages to the pestilential heat of the African coast. To prevent mutinies, slaving captains tried to mix slaves from a variety of tribes and languages, so ships often lay at anchor for months while human cargo was assembled:

> Beware and take care
> Of the Bight of Benin;
> For one that comes out,
> There are forty go in.

This bit of doggerel exaggerates a bit but many men died of tropical diseases while waiting off the coast. Lowther, frustrated by a long layover, led a mutiny aboard the *Gambia Castle* and was elected captain of the ship, which later appeared in the Caribbean in search of plunder. Records show that almost 72 percent of the 110 *Gambia Castle* crewmen who refused to turn pirate and remained behind died within a year. Despite large numbers of rovers who died on the gallows of "hempen fever," more pirates died of drink and disease than ever were hanged; and more perished in shipwrecks than were killed in battle.

Lowther and Low, Two of a Nasty Kind The Cayman Islands, where the buccaneers had once gone for turtles, became something of a pirate sanctuary, and it was there that Lowther met up with Low, whom he made his lieutenant. They were two of a kind and together cruised the West Indies and the Virginia capes, plundering ships and "barbarously using" their captives. Lowther's specialty was putting slow-burning matches between a man's fingers, letting them burn through to the bone if his victim didn't divulge where his valuables were hidden.

In May 1722, after Low split off, taking forty-four pirates to form his own company, Lowther's luck took a turn for the worse. His ship was all but destroyed when an English merchantman put up a surprisingly strong defense when attacked off South Carolina. Lowther was forced to hole up for the winter ' in a remote North Carolina inlet, living off the land while the ship was repaired.

In the spring he headed north and spent the summer plundering vessels with modest cargoes on the Newfoundland Banks. Following the traditional pattern of wintering in a more hospitable climate, Lowther made for the West Indies in August 1723. He seized a few prizes among the islands, but most of the time his men were on half rations for lack of provisions. When it was time for the periodic careening of his sloop, Lowther headed for the small island of Blanquilla, northeast of Tortuga. As their ship lay careened on its side, masts and rigging spread on the beach, Lowther and his company were surprised by the *Eagle*, a sloop out of Barbados belonging to the South Sea Company.

In the warm waters of the Caribbean, marine growth quickly fouled a ship's hull, making it necessary to frequently careen a vessel, and pirates shared their knowledge of many secluded coves that

Figure 5.7: *A painting by an unknown artist showing a British warship chasing a pirate lugger in the English Channel. The three-masted lugger is flying the black pirate flag or Jolly Roger at her mizzen mast. Pirate ships often carried a range of flags, and it was common practice to fly a friendly or neutral flag when approaching a victim, and to replace the flag with the Jolly Roger at the last moment in order to take the victim by surprise.*

were safe. This was one of the surprisingly few cases where pirates were attacked while most vulnerable, while careening their vessel. The *Eagle's* crew engaged the pirates, killing several and capturing all the rest save Lowther, three others, and a small drummer boy. George Lowther was found dead on the beach next to his gun after the *Eagle* had sailed away with his ship and captive men, most of whom were condemned to death by a Vice-Admiralty Court at St. Kitts and executed. He appears to have committed suicide, an act exceedingly rare among pirates.

His former partner fared better, leading his company of sadists on many daring attacks, sometimes cutting rich prizes right out of guarded harbors. He swept over the seas, from the West Indies to Newfoundland and across to the Azores, the Canaries, and Cape Verde. Sailors' blood curdled at the sight of his ensign. Low's appearance was equally bloodcurdling: his face had been disfigured when one of his drunken men missed a prisoner with his cutlass and slashed the side of Low's face open.

Figure 5.8: *George Lowther and his men with a tent made of ship's sails to serve as a temporary shelter. They have beached their ship,* Happy Delivery, *for careening, a process involving hauling the ship onto its side in order to burn and scrape the weeds and barnacles from the ship's bottom, recaulk the seams, and replace rotten or infested planking. The task left pirates temporarily vulnerable.*

Low and his barbaric gang lopped heads off with abandon. They cut off a New England whaling captain's ears and made him eat them seasoned with pepper and salt. They ripped men open for sport, and strung two Portuguese friars to a yardarm and left them dangling. Captain Johnson wrote that they "almost as often murdered a man from the excess of good humour as out of passion and resentment.... for danger lurked in their very smiles."

Whether the lunatic Low ever got his just desserts is unclear. Accounts vary. Some say he perished with all hands when his ship, the *Merry Christmas*, went down, but he may have sailed to Brazil. Many believed that he killed his quartermaster during one of his violent rages and was cast adrift in an open unprovisioned boat with three companions.

New Providence: "Nest of Pyrates"

The pirates of this extraordinary period could not have operated without their bases. None was as important as the Bahamas, particularly the island of New Providence, which by 1716 had become the "Nest of Pyrates" that Virginia's Governor Spotswood was so concerned about. The sparsely settled Bahamas had been a pirate rendezvous since the 1680s, when the obliging Governor Robert Clarke issued privateering commissions, giving them a veneer of legality.

New Providence boasted a perfect pirate harbor, spacious enough for five hundred pirate craft, yet too shallow for pursuing warships; an abundance of provisions; and a good location. Rude taverns sprang up along the beach. Whores and outcasts, mangy dogs, and multiplying rats added to the fluctuating population. Merchants and traders were drawn to the settlement. They catered to the outlaws' nee'ds and purchased their plunder, much of which was then smuggled to the colonies for resale. By 1700 it was reported from Virginia that "all the news of America is the swarming of pirates not only on these coasts but all the West Indies over, which doth ruin trade ten times worse than a war."

A PIRATE REPUBLIC

A number of administrators were sent to New Providence, but the burgeoning pirate forces sent them packing. English pirate chieftains Thomas Barrow and Benjamin Hornigold proclaimed a pirate republic, with themselves as governors of New Providence. They were joined by leading captains such as Charles Vane, Thomas Burgess, Calico Jack Rackham, and Blackbeard. Ex-privateers and outlaws from all over the New World swelled the population of the pirate sanctuary. Logwood cutters from the Central American coast, who had been chased away by the Spanish, came. Escaped indentured servants, male and female, also found their way to the squalid tent city that grew up around Nassau harbor.

After a cruise, pirates looked forward to shore leave on New Providence, which met the recreational requirements of women and wine. It was said every pirate's wish was to find himself not in heaven after death but back on that island paradise where the resting rovers could laze in their hammocks beneath the palms, swinging gently in the fanning breezes. There were whores aplenty, continuous gambling, the camaraderie of fellow rovers, and unlimited drink.

PIRATE FARE AND PIRATE FASHION

Pirate drinks were hearty and spicy. Popular bombo or bumboo, for instance, was a concoction of rum, water, sugar, and nutmeg. Another favorite was a blend of raw eggs, sugar, sherry, gin, and beer called Rumfustian, although it contained no rum. The rovers also greatly appreciated brandy, sherry, and port, which they looted from prizes. Alcoholism was an occupational hazard and led to many untimely deaths.

Pirates liked their victuals hearty and spicy too. The shanty taverns of New Providence offered palate-pleasing dishes of "Solomon Grundy" or *salamagundi,* a sort of spicy chefs salad that included whatever was handy. Bits of meat, fish, turtle, and shellfish were marinated in a mixture of herbs, palm hearts, garlic, spiced wine, and oil, and then served with hard-boiled eggs and pickled onions, cabbage, grapes, and olives.

Figure 5.9: *Captain Edward Low, onetime partner of Lowther, stands at the shore in stormy weather while a ship founders in the heavy seas. Low acquired an evil reputation because of the savage cruelty with which he and his men treated their victims.*

Figure 5.10: A *romanticized engraving by the French artist Auguste-Frangois Biard showing pirates cunningly dressed up as women and harmless civilians in order to deceive an American ship of their intentions as they come alongside.*

Pirates generally wore trousers and jerkins of rough sailcloth at sea, except during an engagement when most donned garments coated with pitch (which could deflect sword thrusts), or doublets of thick leather. On shore many pirates made up for their drab work garb by affecting the colorful dress of gentlemen and exaggerating their mannerisms if not their hygiene. They minced along in silver-buckled high heels, tricorn hats under their arms, clad in plundered combinations of rich-hued and often mismatched garments of embroidered silks and satins, velvets and lace, which often verged on the ludicrous. Some even wore powdered wigs and powdered their stubbled faces like London dandies. Pirates loved gaudy jewelry, wearing rings, elaborate ear pendants, pearls, ornate heavy gold chains, and diamond and emerald crosses stolen from Catholic ships.

SUNKEN TREASURE

In 1715 all twelve galleons of the annual Spanish plate fleet sailing from Havana to Spain sank off the coast of Florida in a hurricane with more than fourteen million pesos worth of treasure. The viceroy of Havana immediately dispatched salvage ships with divers and soldiers to protect them. Working from a camp on shore, they recovered much of the sunken treasure.

Before long every would-be treasure hunter in the Caribbean had heard of the disaster. Henry Jennings, an ex-privateer, raised a contingent of three hundred pirates from New Providence to raid the salvage camp. Their numbers panicked the guards, who fled into the bush, abandoning 350,000 pieces of eight plus assorted other treasure. The pirates then seized a Spanish cargo ship, plundering her rich cargo of cash, cochineal, and indigo, and, based at New Providence, went on to wreak havoc throughout the West Indies.

By 1716 the "Flying Gang," as the New Providence exponents of sea robbery called themselves, had such a stranglehold on the sea lanes from Nova Scotia to the Spanish Main that the governor of Antigua wrote, "I do not think it advisable to go from hence except upon an extraordinary occasion, not knowing but that I may be intercepted by the pirates." Merchant ships traveled in convoys protected by navy warships that charged as much as 12 1/2 percent of a cargo's value for their services.

BELLAMY, "THE ORATOR"

Among the New Providence sea vultures who cruised off the North American coast during this period was Captain Samuel Bellamy, whose ship the *Whydah Galley* was wrecked in stormy seas off Cape Cod while the drunken crew were sampling some prize Madeira wine. In 1984 the remains of the wreck were located in about twenty feet (6m) of water. Salvors brought up several thousand gold and silver coins, as well as gold bars and jewelry, and the ship's bell.

Bellamy left a wife and children near Canterbury in order to cut his teeth in the West Indies as a "wracker," one of those who set out false beacons to lure unwary vessels to their shores and plunder their cargoes. Before cruising the east coast in 1717 Bellamy had "fished" on the 1715 galleons and made a name for himself in the islands as a pirate. His flagship, the *Whydah Galley,* mounted twenty-eight guns. His motley crew of two hundred was predominantly English and Irish with a sprinkling of other nationalities, including twenty-five slaves stolen from a Guinea ship. He was accompanied on his North American cruise by a twelve-gun sloop with a crew of forty, captained by Paul Williams, a mulatto in a periwig. They took prize after prize. One of his crew later testified they had taken fifty vessels. Some ships they scuttled and burned, others were plundered and let go after a certain number of their men had signed articles.

Bellamy epitomized the anarchic spirit of the classic pirate who was at odds with restrictive society, and was dubbed "the Orator" because he enjoyed delivering impassioned speeches. He harangued the captain of a Boston merchant ship taken off South Carolina: "I am a free prince and I have as much authority to make war on the whole World as he who has a hundred sail of ships at sea and 100,000 men in the field."

WOODES ROGERS SAILS IN

Growing alarm about the havoc the pirates were wreaking caused King George I to act. In July 1718 Woodes Rogers, ex-privateer and circumnavigator, arrived at Nassau harbor, where more than two hundred ships rode at anchor and where an estimated thousand pirates were ashore. He brought with him a royal commission as "Captain-General and Governor-in-Chief in and over Our Bahama

Islands." He also had with him an Act of Grace, a royal pardon for all pirates who turned themselves in before September 5, 1718, and were willing to swear an oath to abstain from further piracies. After that date all were to be hunted down and hanged.

Many of the pirates elected to retire from "the sweet trade." A few decamped rather than forswear their seaborne profession, and a small group of dissenters vowed to resist the governor's authority. Charles Vane, their ringleader, turned a French prize into a fire ship fully loaded with ignited explosives. Cut adrift, it bore down on the newly arrived English vessels. The fire's heat set off round after round of cannonballs, musketballs, and bits and pieces of assorted metal scrap. As the powder magazines ignited, the French ship went up like a dazzling, deafening fireworks display, lighting the tropic night for miles around.

In the general confusion Vane, with a number of his men, slipped out to the open sea in a sloop piled high with booty and vanished into the night. For the next three years he remained at large, poaching for silver bullion on the 1715 wrecks off Florida and cruising up the American coast, where he terrorized shipping off the Garolinas. He met Blackbeard ofishore one time and the two crews retired to an inlet for a drunken orgy.

Figure 5.11: *A famous engraving of Edward Teach, better known as Blackbeard, from an early edition of Captain Johnson's famous work. Heavily armed, he sports smoking fuses stuck tinder his hat. Anchored in the bay behind him is his forty-gun ship,* Queen Anne's Revenge, *in which he cruised the Caribbean and eastern seaboard of the United States.*

Eventually his luck ran out. He was shipwrecked and picked up by an English vessel, identified, taken to Port Royal, tried, and then hanged.

Woodes Rogers succeeded in wiping out the pirate nest and even turned some of the men into farmers. Many of New Providence's former denizens found life under the new regime far too disciplined and dull for "men o' spirit" who preferred bumboo to Bible tracts, and they left to return to their old ways. Several former pirates, including Captain Benjamin Hornigold, became Rogers's most trusted agents and went after the backsliders.

BLACKBEARD

One New Providence alumnus who never for a moment considered accepting a pardon was Blackbeard. In the *Queen Anne's Revenge,* he prowled the West Indies with a crew of three hundred men. Engagements like the one in which he bested a British warship made Blackbeard the most talked-about man in the Caribbean. Even the reformed Hornigold had boasted of his role as the great pirate's tutor, and former rovers talked wistfully in the taverns of New Providence of their former comrade's raw courage in taking on and beating a Royal Navy ship. Even in absentia, Blackbeard made life difficult for Rogers, whose hold on his domesticated pirates was tenuous at best.

Blackbeard, a protégé of Hornigold, was not as savage as some pirates but none surpassed him at self-promotion, so that his name lives on as the pirate "whose name was a Terror" and "the Spawn of the Devil." Like so many of his kind, his origins are obscure. His name has been rendered a dozen ways from Teach to Tash to Thatch. He was a native of Bristol, England, and had come to Jamaica first as a deckhand aboard a privateer during the war.

He served his pirate apprenticeship on Hornigold's brigantine *Ranger* and proved so' capable and fearless that his mentor put him in charge of a six-gun sloop with a crew of seventy. The two cruised together, taking prizes off Cuba and elsewhere along the way to the American coast, where they careened their ships on the Virginia shore. After the vessels had been scraped, sealed with tallow, and repaired, they set out again along the coast before heading to the West Indies with their holds brimming with prize goods.

Off St. Vincent they took a French guineaman en route from Africa to Martinique. The cargo of slaves, gold dust, bullion, plate, jewels, and other choice goods was beyond their dreams. Hornigold rewarded Teach's ability by giving him command of the prize.

She was a big Dutch-built ship, strong, and well armed. Blackbeard converted her to his use and patriotically renamed her the *Queen Anne's Revenge.*

During an eighteen-month rampage Blackbeard ranged from Virginia to Honduras, terrorizing shipping and taking at least twenty prizes. He burned some ships but added others to his growing fleet. Most of the American colonies had turned their backs on the pirates, but struggling North Carolina, lacking the lucrative trade in rice and indigo that made neighboring colonies strong enough to shun traffic with smugglers and pirates, still welcomed them.

Figure 5.12: *This evocative picture—-entitled This,* Lean Straight Rover Looked the Part of a Competent Soldier— *was painted by Frank Schoonover and shows Blackbeard's men marching through Charleston, South Carolina. The most remarkable feat in Blackbeard's pirate career was his blockade of Charleston in 1718. He sailed into the harbor with a warship and three smaller vessels and held the town to ransom. Tired of his depredations, Governor Spotswood sent two Royal Navy vessels into Ocracoke Inlet to put an end to his activities.*

Blackbeard was a man of imposing stature and frightening aspect. A mane of thick black hair and a long beard, both plaited with colored ribbons, framed his naturally scowling face with its wild-looking deep-set eyes. In battle, armed with three brace of pistols slung in bandoleers, he appeared to be a fury from hell, with smoldering gunner's matches sticking out from under his hat. Blackbeard's awesome figure was matched by his extravagant and impetuous temperament. Even his own men were subjected to terrifying displays calculated to cow the observers.

One time he was drinking a rousing concoction of rum and gunpowder with his men aboard ship. He was smitten with the idea

Figure 5.13: Blackbeard's Last Fight, *by Howard Pyle. As befitted the image he had created for himself, Blackbeard died in an epic manner. Tracked down and cornered in Ocracoke Inlet by a two-ship naval expeditionary force led by Lieutenant Maynard of HMS* Pearl, *Blackbeard met them head-on in a violent frenzy that culminated in the boarding of Maynard's sloop, and a vicious hand-to-hand encounter ensued. Blackbeard picked out Maynard and shot at him, but missed. Maynard returned fire and hit his quarry, then the two closed with swords. Maynard's sword broke and the fatal blow that came for him missed when another sailor slashed Blackbeard's throat and thus ruined his thrust. Suffering five gunshot wounds and twenty sword wounds, Blackbeard fought on, gurgling blood, until he sank slowly to the deck, unable to function as the life oozed from him.*

of creating a hell of their own to see what it was like. He took two or three obliging crewmen into the hold, where they sat on ballast rocks. Blackbeard had lit pots of brimstone handed down. Then the hatches were closed and the pirates sat in the pitch-black hold breathing the suffocating fumes until one by one they begged for release from the sulfurous prison. Blackbeard finally opened the hatches, delighted that he had held out the longest.

One night he was drinking with his gunner, Israel Hands (the prototype for Stevenson's Israel Hands in *Treasure Island*) and another pirate in his cabin. Suddenly, without a word of warning, he drew and cocked a pair of his pistols under the table. One man, wary of his captain's unpredictable temper, went on deck, while Hands stayed to drink. Blackbeard snuffed the candle and fired both pistols. One slug ripped through the gunner's knee, crippling him for life. When his crew asked why he had done such a thing, Blackbeard cursed them and replied that if he didn't kill one of his men now and then they would forget who he was.

In January 1718 Blackbeard and his crew surrendered to Governor Charles Eden at the town of Bath, on the Pamlico River in North Carolina, under the latest Act of Grace. Eden took a percentage of Blackbeard's booty and made no move to stop him from brazenly careening in the vicinity while he prepared for another cruise. Tobias Knight, the colony's secretary and collector of customs, openly aided Blackbeard before he sailed off for another foray to the Gulf of Honduras.

Stede Bonnet: the Amateur Pirate.

On his first voyage to Honduras, Blackbeard had met up with one of the Golden Age's least likely pirate captains, Major Stede Bonnet from Barbados. Bonnet was a middle-aged, upper-class Englishman who had retired from the army to run a large sugar plantation. It came as a shock to Barbados society when the major deserted his comfortable life for an outlaw career at sea. Discord with his shrewish wife may have spurred him on. In any case he outfitted his sloop the *Revenge* with ten guns, breaking pirate precedent by buying the ship instead of seizing it. He crewed her with seventy men, only some of them seasoned rovers, and took off.

By the time he met Blackbeard, Bonnet had managed to capture a few prizes without being very capable or inspiring confidence in his crew. An odder couple would be hard to imagine: the flamboyant Blackbeard, larger than life with his wild mane and blazing eyes, and the pudgy little dandy with his satin waistcoat, snow-white breeches, and powdered wig. Despite their differences they got on famously and decided to cruise together. Almost immediately Blackbeard realized that Bonnet was a bungling amateur.

The crew of the *Revenge* was pleased when Blackbeard placed his second-in-command, Lieutenant Richards, in charge of Bonnet's ship. He showed uncharacteristic tact in suggesting to Bonnet that "as he had not been used to the fatigues and care of such a post, it would be better for him to decline it and live easy, at his pleasure, in such a ship as his, where he should not be obliged to perform duty, but follow his own inclinations." In effect, Bonnet was Blackbeard's prisoner.

His crew was dead set against him and he had begun to wonder if he wouldn't rather settle ashore in a Spanish colony. There was no getting away, however, as Blackbeard's fleet beat the seas between the Garolinas and the West Indies, taking a dozen prizes.

Bonnet had sought a privateer's commission from the governor while the pirate fleet put in at Topsail Inlet. During his absence, Blackbeard stripped his sloop clean and sailed off, marooning twenty-five of the crew on a sand spit. Bonnet started after Blackbeard, but when he realized he couldn't catch up with him, he cruised up the coast of Virginia into Delaware Bay. He had become more adept during his sojourn with Blackbeard and seized several prizes, including a large ship that he took ashore near the Cape Fear River for modifications.

Figure 5.14: *The last moments of a pirate at the aptly named Execution Dock, also known as Wapping Old Stairs, on The banks of the Thames at Wapping near the Tower of London. The prison chaplain is listening to the last words of the condemned man, while marshals on horseback (one on the left holds the Admiralty oar) keep back the crowd of spectators who always gathered for such an event, eager to witness the manner in which the condemned took his leave. The engraving is taken from a picture by the English artist Robert Dodd.*

It was there that he was captured by Colonel William Rhett of South Carolina. A bribe of gold bought him escape from the Charleston jail, but Rhett caught the portly pirate again and he was brought to trial in Charleston with thirty-three other pirates. Bonnet was found guilty despite his pleas that he had never taken a ship save in the company of "Captain Thatch" (Blackbeard). His ineptitude cost the lives of twenty-eight of his men, many of them young, who were hanged on November 8, 1718.

The gentleman pirate thought he would be spared and wrote a groveling letter begging for mercy from the governor. If reprieved he would separate all his limbs from his body, he promised, "only reserving the Use of my Tongue to call continually on, and pray to the Lord." But South Carolina had had its fill of pirates and he was hanged on December 10 at White Point, near Charleston, the traditional bouquet clutched in his chained hands.

BLACKBEARD'S NEMESIS

Unhindered maritime commerce was vital to Virginia, and Blackbeard's continued haunting of the sea-lanes between the West Indies and the mainland was intolerable. When Governor Eden took no action against the pirate, the traders and plantation owners of Virginia implored Governor Spotswood, implacable pirate foe, to do something. When the Virginia legislature quibbled over how to fund a naval expedition, Spotswood paid the cost himself.

Blackbeard's reign of terror ended a month before Bonnet's execution, when Governor Spotswood sent Lieutenant Robert Maynard of the HMS *Pearl* to beard the pirate in his den. The pirates' sloop *Adventure* and a merchantman prize were at anchor up the Ocracoke Inlet, North Carolina, where word had it Blackbeard planned to carve out a pirate stronghold. He had been smitten by a sixteen-year-old planter's daughter in nearby Bath. He had a wench in every port and is credited with fourteen wives, although the first thirteen were probably amours of the moment who went through mock weddings conducted by one of the pirate officers. The fourteenth marriage was the real thing performed by Governor Eden himself.

Lieutenant Maynard set out in command of two hired sloops, both shallow draft vessels able to navigate the shoals and channels around Ocracoke. He located the pirate ships at dusk of November 21, 1718. Blackbeard and the eighteen men who were with him spent the night getting drunk instead of preparing for battle. Next morning the pirates, still drinking, met the Royal Navy in one of history's great pirate engagements, which ended with a miraculous British victory after the pirates had all but won, having slain the captain of one of the navy sloops and wounded most of the sixty-two sailors.

Blackbeard shot at Maynard from almost point-blank range and missed. The lieutenant's answering shot found its mark. Wounded, Blackbeard fought on like a zombie amid the smoke and stench.

The howling pirate attacked Maynard with his cutlass. Their blades clanged and flashed on the blood-slicked deck until Maynard's broke off near the hilt. Just as Blackbeard went to finish Maynard, a navy man slashed his throat open so that Blackbeard's thrust was deflected, barely grazing Maynard's hand. The dying giant fought on like a cornered wild beast. Drenched in blood that bubbled out of his neck, Blackbeard fired his pistols until the repeated sword thrusts of the sailors who closed in on him finally took their toll.

Blackbeard's grisly head was hung below Maynard's bowsprit as proof of his death and taken to Bath and then back to Virginia, where it was stripped of flesh and hung from a pole at the mouth of the Hampton River. Thirteen of the surviving fifteen pirates from the pirate band were tried and condemned to death at a trial in Williamsburg.

HARSH MEASURES

A spate of executions in Virginia and South Carolina, where forty-four pirates were hanged in one month, broke the pirates' stranglehold on the American coast, but ships in the West Indies were still not safe from rebellious New Providencers. Although he had no legal powers to do so, Rogers held a swift trial for ten pirates Hornigold had brought back to Nassau. Their mutiny threatened to undermine the fragile structure of his authority. Mine were found guilty and one spared because he proved he had been forced to pirate.

On December 12, 1718, two days after they were sentenced, a single gallows was erected with nine nooses dangling over a scaffold resting on three barrels. The pirates were led out and unshackled; after their hands had been tied in front of them, they mounted the gallows. Each of the condemned men was allowed to address the crowd of about three hundred former and potential pirates. The militia of 100 soldiers and irregulars was present to maintain order because feelings ran high in favor of the pirates, who met their end with the courage that had been the hallmark of their careers.

Rogers took a risk executing the pirates, but it paid off. When one member of the sullen mob jumped on a barrel to exhort the others to riot, Rogers calmly shot him dead. After that, although there were mutters of rebellion, Rogers never again had a mutiny. In fact, he had the colonists staunchly behind him when 1,300 Spanish troops poured off four warships to attack the colony in late February of 1720. One British infantry division and five hundred rum-soaked ex-rovers beat off the Spanish.

ANNE BONNY AND MARY READ

Surely the most unusual pirates ever to have set foot on the pirates' island were Anne Bonny and Mary Read, who sailed with Vane's former quartermaster John Rackham, the man whose striped pants earned him the nickname of "Calico Jack." The eighteenth-century press had a field day when the two were brought to trial in Port Royal. They were the subject of numerous ballads.

Anne Bonny was the bastard daughter of William Cormac, a prominent Cork attorney, and a housemaid. The trio settled in Charleston, South Carolin, where Cormac prospered as a businessman and planter. Anne had a wild temper. When she was thirteen she allegedly stabbed a servant girl in the belly with a table knife. When her mother died she took on the duties of housekeeper for her father, whose wealth made her an attractive catch despite her temper.

She eloped with a feckless sailor named James Bonny and Gormac disinherited her. The couple found their way to New Providence, where Anne tired of her husband, who had turned spy for Woodes

Figure 5.15: *Having fought and fatally wounded a fellow pirate in a duel, Mary Read pulls open her shirt to reveal that she is a woman. There were very few female pirates, but the lives of Mary Read and Ann Bonny are well documented. Witnesses at their trial in Jamaica said that they cursed and put up more of a fight than the other members of Calico Jack's crew.*

Rogers. She fell for dashing "Calico Jack." She and Jack stole a sloop at anchor in the harbor and set off "on the account," putting together a crew and taking several prizes.

They captured a Dutch vessel and pressed a number of its crew to sign articles with them. One of the new pirates was a delicately handsome young boy to whom Anne took a fancy. Disappointment may have equaled shock when she discovered her favorite to be a young Englishwoman named Mary Read. Unlike Anne, Mary had spent most of her life in men's clothing. She was the illegitimate child of a London woman whose husband had been at sea for more than a year when she delivered. Mary was apprenticed as footboy to a French lady, but her nature demanded more excitement and she ran away to sea, signing on as a cabin boy on a warship.

Later she served as an infantryman and then in the mounted cavalry in Flanders, where the English and French were

Figure 5.16: *Ann Bonny as depicted in an engraving from an early edition of Captain Johnson's book. After an adventurous early life, Bonny ended up on a pirate ship with Mary Read. With John Rackham they cruised the West Indies until captured off the west coast of Jamaica. They were put on trial at the Admiralty Court in Spanishtown presided over by Sir Nicholas Lawes, the governor of the island, and were only saved from the gallows by revealing that they were both pregnant.*

fighting the War of the Spanish Succession. Mary married but was soon widowed and returned to the life she preferred, signing on a Dutch merchantman bound for the West Indies, which was the same ship Rackham took. Mary shared her secret with Anne, who told Jack. He invited her to join the crew as a full-fledged member. The two women were first-rate pirates, never shirking battle. None among the crew "were more resolute, or ready to Board or undertake any Thing that was hazardous," Captain Johnson asserts.

Their ship was anchored off Jamaica's north coast in late October 1720 when they were challenged by a privateer sloop, whose captain had a commission from the governor of Jamaica to take pirates. There was a short, sharp action that ended with the pirates' surrender. Captain Jonathan Barnett, the sloop's commander, testified in court at Spanishtown, Jamaica, that only two of the pirates had put up any fight and they had fought like wildcats, using pistols, cutlasses, and boarding axes before being overpowered. One of the two had fired into the hold where the men were hiding, screaming like a banshee that they should come up and fight like men. When it was discovered that these two pirates were women, it was arranged that they be tried separately from the men.

Anne and Mary were sentenced to hang along with Rackham and the rest of those who had been convicted. The courtroom was astounded at the women's answer to the judge's routine inquiry as to whether any of the condemned had anything further to say. "Milord," came the reply, "we plead our bellies." By law the court could not take the life of an unborn child by executing the mother. An examination revealed that the two pirates were indeed both pregnant. Mary died of fever in prison before her baby was born. Anne disappeared, perhaps paroled through her father's influence. She fades from history in a fury, for as Jack Rackham was going to be hanged (at a place near Port Royal still known as Rackham's Cay), he asked to see her. She gave him a scornful look and spat qut that, "Had you fought like a man, you need not have been hang'd like a dog."

THE END OF AN ERA

A number of factors brought the Golden Age of Piracy to a gradual conclusion. By the end of the 1720s the naval abuses mentioned earlier had ended, and as the Crown ceased issuing proprietary charters the quality of government officials improved. Corrupt officials were replaced with better men, and the threat of revocation encouraged administrators to suppress piracy in those colonies that retained their charters.

The expanded Piracy Act of 1721 came down hard on those who "shall trade with, by truck, barter, or exchange," with a pirate, making them equally guilty of piracy. This legislation greatly aided officials like Governor Spotswood, who had been frustrated by the extent to which the community was involved with pirates. With time local economies were strong enough to shun trafficking in pirated goods.

In addition, the barbarous behavior of so many of the sea outlaws sickened people, outweighing their romantic appeal. Lowther and Low were the epitome of pirate savagery, but atrocities were commonplace and escalated as suppression and hanging of pirates increased. The New England-based John Roberts, for example, invariably tortured captives, urging his men on to the most sadistic acts. They whipped men to death, used them as target practice, and sliced off ears and noses. Captured pirates faced torture themselves. In 1725 Scots pirate Captain John Gow refused to plead at his trial at Newgate and was ordered pressed to death. Faced with the prospect of being crushed fo death, the only torture then allowed by law, Gow decided to plead not guilty. Found guilty, he was hanged and his body taken to the shore of the Thames at Greenwich to hang in chains as a warning to others.

THE "GREAT PYRATE ROBERTS"

Bartholomew Roberts, the last of the Golden Age pirate captains, was also the greatest. In his own day he was the "Great Pyrate Roberts," undisputed king of rovers; fearless, original, and a superlative seaman. He was born in Pembrokeshire and, according to Captain Johnson, had the dark complexion common among Welshmen. "Black Bart" was a handsome, commanding figure with a taste for elegant dress. He died during a battle wearing a rich crimson damask waistcoat and breeches, a scarlet plumed hat, and a massive gold chain attached to a jeweled cross.

During engagements he sported two brace of pistols tucked in a silk sling over his shoulder in addition to the razor-sharp cutlass that was his preferred weapon. He kept a tight rein on his crew, forbidding gambling for money and encouraging prayer. Roberts was piracy's only recorded teetotaler, but the fact he preferred tea to rum didn't mean he was a namby-pamby.

Roberts's amazing boldness inspired his men to the most daring acts. He crisscrossed the globe with a fleet so formidable that naval squadrons in the West Indies sent out to capture him turned away at the mere sight of his flag rather than risk confrontation. Even with a single ship, Roberts performed amazing feats. In June 1720, for example, with a single ten-gun sloop and a crew of sixty, he took twenty-two ships lying at anchor in Newfoundland's Trepassey Bay.

More than 1,200 men were aboard the vessels when he appeared. With drums beating and trumpets blaring, the pirates proceeded to take every one as the terror-stricken crews piled into launches and pulled for shore. The pirates plundered and sank all but a brigantine, which was needed to carry the booty. Roberts wasn't fazed a bit by a large French flotilla he encountered not far from the harbor. He attacked and destroyed the entire flotilla, keeping only a large brigantine that he renamed the *Royal Fortune* and made his flagship.

According to Captain Johnson, he embarked on a piratical career not because he had no other way of making a living, but "to get rid of the disagreeable Superiority of some Masters he was acquainted with ... and the Love of Novelty and Change." He turned pirate at the unusually advanced age of thirty-six. Born in 1682, he began his career in the merchant service and in 1719 he shipped as second mate on the *Princess* bound from London to the Guinea Coast to load slaves. Off the West African coast the *Princess* was seized by the famed Welsh pirate, Captain Howell Davis.

HOWELL DAVIS, BLACK BART'S TUTOR

Davis's pirate career began when the slaver on which he was a mate was captured by Edward England, the Irish pirate, who was en route from Nassau to Madagascar. After prowling the West Indies, Davis headed for the eastern seas like so many of the West Indies rovers.

By the time Davis met Roberts he had taken many prize cargoes of gold dust, gold bars, coins, ivory, and slaves. Davis took pride in never having to force a seaman and was delighted when his compatriot Roberts was willing to sail with him.

Six weeks later, after taking several prizes, including a Holland-bound ship with a rich cargo and the governor of Accra aboard, Davis was killed in an ambush at the Portuguese settlement on Prince's Island in the Guinea Gulf.

ROBERTS TAKES CHARGE

Despite his brief tenure, Bart Roberts's valor and intelligence had impressed his pirate mates so much that they elected him their new captain. Roberts retaliated for Davis's death by leveling the Portuguese settlement. Still in a vengeful mood, he then crossed the Atlantic to Bahia, the Bay of All Saints in Brazil. There he found forty-two Portuguese ships at anchor loaded for the passage to Lisbon. He singled out the biggest, the vice-admiral's forty-gun vessel, and sailed up to her so brazenly that the pirates were aboard before the

Figure 5.17: *A silver-mounted presentation pistol, c. 1730, made in London and presented to Captain Reed of the privateer* Oliver Cromwell *by the Council of the West Indian island of St. Christophers. This is typical of the sort of weapon that would have been used by pirates in the Caribbean during the Golden Age. The butt was designed to be used as a club in hand-to-hand fighting.*

Figure 5.18: *Bartholomew Roberts at Whydah on the Guinea Coast of Africa in 1722 with two of his ships in the background, the* Royal Fortune *and the* Great Ranger, *His flagship is flying the distinctive pirate flags he designed: he is depicted standing with each foot on a skull, one labeled ABH, the other AMH. These stood for "A Barbadian's Head" and "A Martinican's Head," respectively—a comment on the efforts of the governors of these islands to put an end to his Caribbean activities. At Whydah, Roberts held eleven ships to ransom (visible here) until they paid in gold for safe passage. Shortly after this incident, Roberts was tracked down and killed during a sea battle off Cape Lopez, as a result of which all his crew were captured.*

Portuguese realized what was happening. Before the two warships assigned to guard the Portuguese convoy could reach him, Roberts had sailed off with his prize containing sugar, hides, and tobacco, as well as forty thousand gold coins valued at about £80,000, gold trinkets, plate, and the diamond-studded cross that afterwards reposed on *his* chest rather than the king of Portugal's, for whom it had been designed.

The pirates sailed north to the cheerless little Spanish colony at Devil's Island off Guiana, where the señoritas proved especially eager to trade for the Portuguese loot. There is a legend that the treasure taken at Bahia is stilt buried deep inside a cave on Little Cayman Island. But it is unlikely the pirates had much left after a couple of weeks of carousing, gambling, and wenching at Devil's Island.

A NIGHTMARE VOYAGE

Roberts took his men into the Caribbean in the spring of 1720 and then north. He spent the summer rampaging around Newfoundland and the New England coast with a complement of about 100 men, including recruited cod-splitters who had been only too happy to leave their back-breaking work for piracy. After a spell in the Caribbean, where naval patrols were numerous, Roberts set sail for West Africa via the Cape Verde Islands. It turned into a nightmare voyage. Once across the Atlantic the prevailing southerly wind forced his ship too far north to reach the Cape Verdes. He had to recross the ocean and then head south into the Caribbean, taking advantage of the western Atlantic's prevailing northeasterly trade winds.

The ship didn't touch land for two thousand miles (3,200km). The days became weeks. Roberts doled out one swallow of water a day from the only remaining barrel of water until it was empty. The pirates were so crazed with thirst that some drank seawater or their own urine. Dysentery and fevers flashed through the crew, killing quite a few. Roberts with his unshakable faith in God's providence was the only man with a spark of hope. Most of the men were beyond caring whether they lived or died when they found themselves suddenly in shallow water, indicating land was not far off. The following day they sighted land and by evening were toasting one another with fresh water from the Maroni River in Suriname on the coast of South America.

DAZZLING FEATS

After this almost fatal episode Roberts ignored the threat of naval patrols and unleashed his lethal forces on the Caribbean in a desperate and dazzlingly successful campaign. The governor of the French Leewards reported that in a four-day period at the end of October 1720, the pirates had "seized, burned or sunk fifteen French and English vessels and one Dutch interloper of forty-two guns at Dominica." Sailing right under shore batteries at the Basseterre Road on St. Kitts, Roberts plundered and burned a number of vessels at anchor. Word spread that the pirates were slicing off the ears of prisoners, lashing them to the mast or yardarm for target practice, whipping them to death, and committing other atrocities.

By the spring of 1721 there was little left to plunder. The holds of Roberts's ships were solidly packed with booty that could be traded at a premium for Guinea gold. So in April he made for Africa, landing on the coast of Senegal and then proceeding to Sierra Leone, where his men careened their ships and relaxed in the company of independent English traders who operated in defiance of the Royal Africa Company's monopoly.

During the summer and early fall the arch-pirate plundered up and down the Guinea coast, sometimes attempting to trade with local tribes and battling them when they refused. One attack around Old Calabar was still a part of the orally transmitted history of the natives as late as the 1920s. At Whydah Roads, where Captain Johnson wrote that there was "commonly the best booty," Roberts took eleven French, English, and Portuguese ships in a single day, releasing each for a ransom of eight pounds' worth of gold dust. One of the receipts he wrote for the gold refers to the pirates as "we Gentlemen of Fortune." No pirate captain ever had complete command of his crew and, when one captain refused to pay the ransom, his ship with its cargo of chained slaves still aboard was set afire against Roberts's orders.

LUCK RUNS OUT FOR ROBERTS

Captain Roberts seemed invincible, but in February of 1722 the *Swallow*, a fifty-gun Royal Navy warship commanded by Captain Chaloner Ogle, caught sight of the pirate squadron at anchor in the lee of Parrot Island off Cape Lopez on the Guinea coast. HMS *Swallow* and HMS *Weymouth* had been hunting for Roberts for six months, frustrated by the coastline's interminable maze of tangled jungle, corkscrew rivers, and hot swampy lagoons. The pirates at first thought the *Swallow* was a Portuguese merchantman and Roberts sent the *Ranger* after her. Ogle lured the *Ranger* out of sight and then turned on her, slaying ten pirates and taking more than 100 others prisoner, many of whom were hideously wounded by an explosion.

When Ogle returned to Parrot Island at dawn five days later, the pirates' ships still rode at anchor. Captain Roberts was breakfasting on *salamagundi* and tea while most of his men slept off prodigious hangovers from liquor taken from a prize the

Figure 5.19: *A portrait of Captain Chaloner Ogle, knighted for his action against Roberts and ultimately to become Admiral of the Fleet. Ogle, in command of HMS* Swallow, *was sent in pursuit of Roberts after the Whydah incident. He found him at anchor under the lee of Cape Lopez. Roberts's crew had just previously taken a prize well stocked with liquor and most of his men were drunk, save Roberts himself. At the onset of the action, Roberts led from the front but was struck by a grapeshot in the throat. His death took the fight out of his men and they surrendered after throwing their leader overboard, according to his request.*

previous day. The *Swallow,* flying the French flag, hove into view without causing alarm, but as it made straight for the *Royal Fortune* one of Roberts's crew, a navy deserter, recognized her as his old ship and raised a cry.

The pirate captain determined to cut cables and run since his crew was not in fighting trim. Then he made a fateful change in plan, ordering the helmsman to steer right for the *Swallow.* As the two ships drew closer, Roberts jumped onto a gun carriage to direct his ship's fire. The *Swallow* launched a broadside that toppled the pirate ship's mizzen topmast. The pirates returned fire.

Figure 5.20: *Cape Coast Castle on the Gold Coast of West Africa provided a fortified base of operations from which British trade and colonial control could be administered from 1664 onward. It became the overseas headquarters of the Royal Africa Company and a center for the trading of ivory and slaves. The castle also served as a military post from which antipiracy activities could be directed. It was at Cape Coast in 1722 that the largest ever trial of pirates was held, involving Bartholomew Roberts's surviving crew.*

When the smoke cleared they saw Roberts's great body slumped on the the deck, his throat ripped open by a blast of grapeshot. His men, many of them sobbing, respected his last wishes and heaved his bleeding corpse overboard in all its finery—jewels, weapons, and all.

Thus ended the stunning career of the "Great Pyrate" Roberts, who in less than four years had captured over four hundred sail. His men fought on, but without their charismatic captain they lost their will. At two in the afternoon 152 survivors, many too drunk to stand, surrendered to a navy boarding party. One befuddled pirate who had spent the entire engagement in a drunken stupor suddenly came to and seeing the *Swallow* nearby shouted: "A prize! A prize!" urging the others to take her. A large quantity of plunder was found aboard the pirate ship, and Captain Ogle, who is the only naval officer ever knighted for capturing pirates, eventually became admiral of the fleet and acquired a fortune from the prizes he had taken off the African coast.

THE TRIALS AT CAPE COAST CASTLE

A total of 264 pirates from the *Ranger* and the *Royal Fortune* were carried to Cape Coast Castle for the largest pirate trial ever held. Nineteen died of their wounds before the trial; some were released. Only eight of the 165 men tried were grizzled veterans. Two had previously sailed with Blackbeard and six were with Davis when he died. The rest were recruits. Seventy-four men including the musicians were acquitted. Fifty-four men were sentenced to death, two of whom were reprieved. Another seventeen were sentenced to prison in London, of whom all but four died en route to England. Not one of the twenty sentenced to seven years' hard labor in the Gold Coast mines of the Royal Africa Company outlived his sentence.

Fifty-two unrepentant pirates ranging in age from nineteen to forty-five were hanged "like dogs" in batches over a two-week period in April 1722. As he was led to the gallows erected outside the castle ramparts, one of Roberts's original crew, "Lord" Symson, saw a woman he knew in the crowd that had gathered to watch the show. She was Elizabeth Trengrove, an Englishwoman who had been a passenger on a ship.Roberts had captured. "I have lain with that bitch three times," he shouted, "and now she has come to see me hanged."

With the mass hangings on the African coast, the curtain came down on the Golden Age of Piracy. The bodies of eighteen of the pirates, coated in tar and wrapped with metal bands, were strung from gibbets on the hills overlooking the harbor. The

Figure 5.21: *One of the actual death sentences passed against nineteen of the accused at Cape Coast in J722. The men were hanged outside the gates and some were left in chains as a warning to others. Ninety-one men were found guilty in total, with fifty-four sentenced to death, fifty-two of whom were hanged.*

macabre spectacle of the rotting cadavers twisting in the wind gave notice that there was no longer any place for the pirates in a world dominated by global European imperialism—a world that, ironically, piracy had helped make possible.

Sporadic episodes of piracy continued to flare up, and even today pirates plague shipping in several far-flung areas; for as Roman historian Dio Cassius noted in the first century A.D., "There was never a time when piracy was not practiced. Nor may it cease to be as long as the nature of mankind remains the same." But never again has piracy been a significant factor in shaping international political and economic policy. Nor has any pirate since captivated the public as did the antiheroes of the Golden Age, whose long shadows reach across the centuries.

CHAPTER SIX

Indian Migrant Women and Plantation Labour in Nineteenth- and Twentieth-Century Jamaica: Gender Perspectives

VERENE A. SHEPHERD

Female labour force participation in the plantation industry of Jamaica and the wider Caribbean has a long history. White indentured servant women worked on the early plantations in the seventeenth century and enslaved African women dominated the plantation field labour in the period of slavery. Indian women were (reluctantly) recruited for plantation labour in Jamaica in the mid-nineteenth century as proprietors searched for a new system of slavery in the aftermath of the abolition of African slavery in 1834, and the subsequent withdrawal of significant numbers of ex-slave workers from the sphere of estate labour. The recruitment of Indian women for commodity production on plantations was not unique to Jamaica as similar patterns of labour recruitment were organized for Fiji, Natal, Mauritius and other territories within the colonized Caribbean (Beall 1991: 89-115; Lal 1983; Laurence 1971:24-46,1994; Look Lai 1993:107-53; Ramachandran 1994: 132-4; Reddock 1984, 1994). Though their numbers were small, Indian women comprised the largest proportion of female immigrant labourers recruited for Jamaican plantations.

The history of plantation labourers of diverse ethnic origins has traditionally been subjected to race and class analysis. The perspective of gender has also now assumed a greater role in the historical discourse of the Caribbean as more and more research reveals that the experience of proletarianization and racial and ethnic oppression was not the same for men and women. In the case of the history of immigrants, as Jo Beall (1991: 89-115) emphasized in her study of indentureship in Natal, a study of the separate experience of Indian women under contract as indentured labourers serves to illustrate how the materialist feminist discourse can point to crucial issues in a broader analysis of relations of gender which can support analyses of class and race and/or ethnic relations.

Accounts of the differential socio-economic positions of minority ethnic groups in late nineteenth- and early twentieth-century Jamaica, for example, usually emphasized the depressed conditions of the Indians. Further examination of the comparative positions of male and female migrant workers usually leads to the conclusion that while it is true that the common history of struggle against migration, racism and the general conditions of indentureship united Indian men and women, as in Natal and elsewhere, Indian female plantation labourers in Jamaica were subjected to ultra-exploitability. Indeed, contrary to Emmer's (1985) conclusions for Suriname that migration and indentureship were vehicles of female Indian emancipation (a claim denied by Hoefte 1987), the indentureship experience in Jamaica did not seem to result in any great degree of social betterment for Indian women, though there were, undeniably, small gains.

This chapter will explore the ways in which gender ideology contributed to the shaping of the social and economic experience of female Indian plantation workers during and after indentureship. Gender clearly played a role in the shaping of socio economic policies and social consciousness in post-slavery Jamaica. Jamaican slave society had been characterized by a paucity of working-class white women, with enslaved black women comprising the majority of field workers. The gender division of labour, while weakly instituted in the area of field labour, was observed in higher status slave occupations. An ideological shift occurred, however, in the post-slavery period. A combination of factors, among which were European patriarchy and Victorian ideals, imported into the Caribbean, dictated that men should function in the 'public' domain of wage labour while women were to inhabit the sphere of uncompensated work in the home. This process had been taking place in Britain since the end of the eighteenth century. By the time of the Indians' arrival, therefore, attempts were already underway to adhere to a 'proper gender order' in the division of labour. This ideological shift had a significant impact on the recruitment of female plantation workers, explains the marked sexual disparity in migration schemes, and partially accounts for the depressed socio-economic life of the contract and post-migrant labourers.

GENDER AND THE RECRUITMENT OF FEMALE PLANTATION WORKERS

In order to understand how gender functioned in the experience of Indian plantation workers in Jamaica, it is first necessary to examine how gender affected migration policies and the numbers and types of women recruited. The tendency in the mid-nineteenth century to dichotomize work and family, public and private, determined the landholders' attitude to the recruitment of Indian women. It is clear that they initially regarded the importation of women as uneconomical. In the first place, planters did not regard Indian women as capable agricultural workers. They believed that Indian men worked more efficiently and productively. Their view, as expressed by the Acting Protector of Immigrants, was that, 'indentured women as a rule are not nearly the equal of the men as agricultural labourers', and in the early twentieth century, when efforts were being made to increase the numbers of women shipped, planters objected to being obliged to pay to import women who they claimed were 'not as good' as male agricultural workers (C.O.R.1843-1917a: 571/1). Second, unlike during slavery when black women had the potential to reproduce the labour force (though fertility rates were generally low), the progeny of Indian females could not automatically be indentured; so Indian women were not initially highly valued for their reproductive capacity. Indian children could only be indentured at age sixteen, though in practice many were used in the fields from age six or so, receiving wages of between 3 and 6 pence per day. But this was only with their parents' consent. Furthermore, proprietors were obliged to provide rations for immigrants' children, whether such children had been imported from India or born in the colony. In some cases, they also had to stand the cost of hiring nurses and establishing creches to look after immigrants' young children (ibid). Third, landholders were not too concerned initially about the social life of the immigrants; so the sexual disparity and its implications for the stability of family life did not preoccupy them. Indeed, the requirement to provide immigrants with return passages at the end of their contracts seemingly made it less critical to be concerned about the construction of the Indian family and the impact of a shortage of women.

Planters were thus not particularly anxious to employ indentured women as part of the immigrant plantation labour force. Lai and McNeil (1915: 313) reported that 'so far as we could ascertain, employers are not particularly anxious that women should work provided that they are properly maintained and absence from work does not merely mean exposure to temptation and the possibility of serious trouble'. Some employers were even prepared to reduce women's indentureship from five to three years; and those who did not support this measure only opposed it because of the pressure of Indian men who could not support non-working partners, and planters who argued that women might have too much free time on their hands; that 'a woman who is not occupied otherwise than in cooking her husband's food is more likely to get into mischief' (Lal and McNeil 1915: 313). But even though five-year contracts remained, in the first three years of their contract, as Lal and McNeil observed, 'a woman who is known to be safely and usefully employed at home will not be sent out to the field' (ibid).

Proprietors, therefore, maintained a gender-specific importation policy which favoured men; and recruiters in India mostly carried out the instructions of the Jamaican planters regarding the composition of recruits. Up to 1882, recruiters were even paid less for each female emigrant recruited. The rate paid was 6 annas per head for females and 8 annas for males (IOR 1882: V/24/1210). The need to increase the numbers of women shipped to the region soon caused an increase in this rate. On the ship *Blundell* in 1845 which carried the first group of Indian indentured workers to Jamaica, women comprised just 11 per cent of the total of 261. When the number of girls under age ten is added to this figure, then the percentage of females increases to 15 percent. On the *Hyderabad* in 1846, women made up 12 per cent of the total shipment of 319 with total females comprising 15 per cent as on the *Blundell*. On the *Success* in 1847, women comprised 10 per cent of the shipment of 223 adults (C.G.R 1845-1916, Ship's Papers).

Jamaican planters relented and adjusted the unfavourable ratio only in the face of governmental pressure to conform to a 40:100 female-male ratio for immigrants over age ten. But they did not go as far as to support a suggestion by the 1913 investigating team of Lal and McNeil of a further increase in the ratio to 50:100 regardless of the age of the female immigrants. This was despite the support given to the suggestion by the delegates appointed to consider the future of indenturesbip after the end of the First World War when certain difficulties in importing labourers were anticipated. They recommended that, 'wherever it is possible to find a sufficient number of females willing to emigrate, this ratio [of .40:100] should be increased [to the level suggested by Lal and McNeil]'(CO.R. 1847-1917a: 571/4).

Planters were also forced to agree to an increase in the importation of women because of the economic imperative of encouraging the settlement of Indians rather than their expensive repatriation. Indeed, by the end of the nineteenth century, planters had successfully influenced changes in immigration policy as it related to the length of contract, repatriation and the period of industrial residence. By that time, contracts had been lengthened to five years, the period of industrial residence extended to ten years after which repatriation could be accessed; and the immigrants were being required to pay a portion of the cost of their return passage. These changes were influenced by the economic downturn after 1884 evidenced, for example, by an increase in the cost of production on estates. In this period also, there was an increase in the cost of immigrants' transportation to and from India. The landowners therefore increasingly clamoured for time-expired Indians to remain in the region and form a permanent labour force instead of opting for repatriation. Permanent settlement, they believed, would compensate for the high cost of importing indentured labourers.

The result of the change in the planters' attitude towards the importation of female immigrants was that by the late nineteenth and early twentieth century more women were being imported. On the *Chetah* in 1880, there were 112 females (nearly 44 per cent of the total) and 256 males. Females comprised 31 per cent of the total number shipped on the *Hereford* in 1885 and 30 per cent on the *Volga* in 1893-4. Of the 2,130 imported on the *Moy, Erne* and *Belgravia* in 1891, females totalled 689 or 32 per cent. On the *Belgravia* which imported 1,050 in all, females numbered 360 compared with 690 males (Shepherd 1994: 50-2; Shepherd 1995:237-8). On some ships in the nineteenth century, the proportion of women landed in Jamaica even exceeded the stipulated female-male ratio of 40:100. For example, in the 1876 shipment the female-male ratio was 46:100 and it was 43:100 in 1877/8. Between 1905 and 1916, the percentage of women on each ship which arrived ranged from 22 to 30 (Shepherd 1995: 238).

Recruiters were not only encouraged to obtain more women, but more women 'of a respectable class', preferably as part of families. This meant excluding single, unaccompanied women. This was because there had developed an erroneous notion in India that single women were forced into prostitution in the colonies. Some visitors to Jamaica even seemed to have shared this belief—however unfounded. One H. Roberts, a noted opponent of immigration, claimed in 1847 that 'the utter disproportion of females in each locality tends greatly to the increase of vice and immorality' (C.O.R. 1847-1917a: 318/173). Lal and McNeil later agreed on this view of the existence of prostitution, though they disagreed that it was widespread. According to their report, 'of the unmarried women, a few lives as prostitutes whether nominally under the protection of a man or not. The majority remain with the man with whom they form an irregular union.' They attributed this to the fact that some women were 'constantly tempted into "abnormal" sexual behaviour by single men with money. But they (the women) are open to temptation as on all estates there are single men who have more money than they need to spend on themselves alone' (see JT, 8 May 19,15).

The Acting Protector of Immigrants in Jamaica also claimed that prostitution was noted among some Indian women in the island; so while agreeing that more women should be recruited, he warned that these should be of a 'better class'. According to him, 'it is no use increasing the proportion of women if they are to be picked up off the streets. They will only lead to further trouble as these women go from man to man and are ceaseless cause of jealousy and quarrels' (C.O.R. 1847—1917a: 571/1). It was in an attempt to induce women of a 'better class' to emigrate that landholders tried to reduce the period of indenture for women to three years. They believed that a shorter indenture and the promise of domestic life thereafter would be attractive inducement for the women and for their husbands. But not only were indentured men unable to afford the cost of maintaining their wives on account of the low wages they received, but Indian women demonstrated a preference for wage labour over uncompensated labour in the home.

Family emigration was supposed to help to solve the 'problem' of the emigration of too many single women. Before the early twentieth century, family emigration had been discouraged on the ground that this necessitated the importation of a large number of children who would increase the risk of epidemics, raise the mortality rate and delay the sailing of ships. This last matter was a perennial cause of concern, judging from the correspondence of the Protector of Emigrants in Calcutta in which he often produced figures to show how much delay could be caused by any unusual illnesses.

The fact that only children aged sixteen years could be indentured and that in many cases women would not emigrate without their children, had also been a deterrent Towards the end of the indenture period, the view was that 'the emigration of whole families will be encouraged' (CO. R. 1847-1917a: 571/6). While by the early twentieth century children of all ages were allowed to emigrate with their families, a preference was expressed for the recruitment of girls who would eventually increase the number of potential wives. Still, the numbers of boys and male infants shipped continued to exceed the numbers of girls and female infants. This was revealed in the sample survey of ships arriving in Jamaica between 1845 and 1916 which showed 508 boys and 333 girls being imported.

There was, predictably, some opposition to the emigration of families, on financial grounds. G. Grindle of the Colonial Office, in response to the recommendation of the Indian government officials, Chimman Lai and James McNeil, stated that 'the encouragement of the emigration of whole families, which in itself is a desirable feature of the scheme, will make the proportion of passages to working emigrants higher than under the existing system, especially as women will be under no obligation to work (C.O.R. 1847-1917a: 571/6). Despite the wishes of the planters, the majority of Indian women emigrated, not as part of a family (to conform to the mid-nineteenth century Victorian ideology of the 'proper gender order'), but as single women who had signed their own contracts for labour on the plantations. On the ship *Indus* which arrived in the island in 1905, for example, only 29 per cent of the women were noted as married and accompanied by spouses; 71

per cent were recorded as single or unattached. On the *Indus* of 1906, 33 per cent of the women were married and 66 per cent were single or unattached. Single Indian females continued to outnumber the married in the post-indentureship period. In the census of 1891, 2,851 out of 4,467 Indian women were recorded as single. Similarly the 1911 census showed 4,467 single Indian women, 2,479 married and 454 widowed. Forty-three women did not state their marital status. It should be stressed though, that there was some under-reporting of the marital status of Indians in the Caribbean in the censuses. Marriages performed according to Hindu or Muslim rites and which were not registered according to the legal requirements of the region, were not recorded.

Planters were also concerned about the tensions which developed among Indian men over scarce Indian women, and the violence against Indian women which resulted; for Indian men did not at first respond to the scarcity of female Indian partners by cohabiting with African-Jamaican women. There were frequent reports from men that Indian women were displaying a great degree of sexual freedom and independence. Some single Indian women reportedly changed partners frequently and seemed unwilling to marry any of the men with whom they developed sexual relations. This behaviour resulted in uxoricide and wounding of Indian women by jealous Indian men. Chimman Lal and James McNeil expressed the view that 'perhaps the best guarantee against infidelity to regular or irregular unions is the birth of children'. But the birthrate among indentured women remained low for the entire period of indentureship (JAR 1920).

In 1913 the Acting Protector of Immigrants in Jamaica supported an increase in the numbers of females shipped to the island as a remedy for the growing incidence of abusive behaviour towards Indian women. In a letter to the Colonial Secretary he reiterated that 'increasing the proportion of women would most likely reduce the number of cases of wounding and murder on account of jealousy, and be an excellent arrangement from the male immigrants' point of view as there would not be such a dearth of East Indian women as there is now on a good many estates' (C.O.R. 1847-1917a: 571/1).

Despite the attempts to increase the numbers of women in the island, the female Indian population in Jamaica was outnumbered by the male Indian population for the period of indentureship, as is indicated by Appendix 1 which is based on the population censuses. It is clear that up to the end of indentureship in 1921, Indian women were still less than 50 per cent of the total Indian population, though the proportion had improved from 31.5 per cent in 1871 to 45.2 percent in 1921.

LOCATION OF FEMALE INDIAN PLANTATION WORKERS

Most Indian women resided in rural Jamaica, reflecting the pattern of location of agricultural units on which they were indentured. The 1861 census detailed the distribution of the population by parish; but did not carry details on ethnic distribution as opposed to parish distribution by 'colour'. However, the details on the 'native country' of the inhabitants enabled an analysis of the distribution of the population from India by parish. It showed the majority living in Westmorland, Metcalfe, Clarendon, St. Mary and Vere (Higman 1980: 31-9). In 1871, the census showed a slightly altered parish distribution. The total Indian population in that year was 7,793, comprising 5,339 males and 2,454 females. The majority of Indian women were settled in Clarendon, Westmorland, St. Thomas-in-the-East, St. Catherine, St. Mary, St. James, Hanover, St. Elizabeth and Portland.

In 1881, the Indian population totalled 11,016, of whom 4,075 were female. The parish distribution in order of numerical importance was: St. Catherine, Westmorland, Clarendon, St. Mary, St Thomas and Portland. Similarly, in the 1891 census, Indian women numbered 4,467 or 1.6 per cent of the total population of Jamaica. The majority were to be found in Westmorland, Clarendon, St. Catherine, St. Mary, St Thomas, Kingston and Portland. From 1911 to 1943, St. Mary was the leader in terms of numbers of Indian females. One reason for the changes noted in the parish distribution pattern was that after 1891 the banana-producing instead of the sugar-growing parishes employed the majority of Indian labourers. On the whole, though, the censuses revealed a distribution which reflected a preponderance of Indian women in the rural areas, whether in the sugar- or banana-producing parishes. This rural bias was most marked up to 1881, and more notable among Indian women than among other female ethnic minorities. The census of 1881, for example, revealed that Chinese and white women—few of whom were involved in agriculture—were concentrated in Kingston and St. Andrew.

A shift took place only after 1943 when the rural to urban migration movement was reflected in the growing numbers of Indian women in the urban centre. This trend had actually started after 1881, a reflection of the trek of ex-indentured immigrants from rural parishes, but it intensified after the 1940s. The parishes experiencing the highest drain were St. Mary and Westmorland (see Appendix 2).

WOMEN, GENDER AND INDENTURESHIP

Indian women began their experience in Jamaica as indentured labourers. One of the consistent features of colonial and imperial organization of migrant labour, the indenture or contract system, provided a means of retaining labour in the short and medium term and an institutional framework to facilitate the further movement of labour in the post-slavery period. At the inception of labour migration, contracts were for only one year, with renewal being optional. The period of contract for men and women was later extended to three years. By 1870, immigrants were given five-year contracts with repatriation due only after a further five years of continuous residence in the island.

Archival records yield more data on issues relating to fertility, the sexual disparity in migration schemes and male-female social relations than on the gender differences in the working condition of immigrants, specifically the extent to which they were subjected to sex-typing of jobs and gender discrimination in wages for equal work. But from the data available, it is clear that female Indians were subjected only to a limited form of the sex-typing of jobs according to which women were confined to service industries and men to agricultural field or factory positions. This sex-typing of jobs under capitalism was one of the forms of the sexual division of labour which European colonizers attempted to replicate in the Caribbean. It was traditionally created by the interaction in capitalistic society between the family and public economic life. But as Indian women could not be confined to the private sphere as wives of indentured men, and as there were insufficient openings for domestic servants in the scaled-down planter households of the post-slavery period, landholders were forced to use them in the fields.

Female Indian plantation workers were placed in one of three gangs on sugar and banana estates. The gangs were headed by males—usually African-Jamaican. As during slavery, placement in gangs was determined by age, physical condition and gender. In the period of indentured labour migration, however, race/ethnicity was added to the other criteria for gang allocation; for invariably African-Jamaican workers were placed in the first gang (Shepherd 1994: 57). Up to 1880, Indian women were indentured on sugar estates. After 1880, they were primarily located on banana plantations. This shift was explained by the post-slavery decline in the sugar industry and the concomitant expansion in the banana sector, some sugar estates being turned into banana plantations. From 113 banana plantations in 1893, for example, the island had 435 by 1910. By the following year, 47 per cent of Indians were located on banana plantations in contrast to 39 per cent on sugar estates and 2 per cent on livestock farms (Shepherd 1994:120).

The proprietors maintained a gender division of labour in non-field occupations. Thus while Indian women were confined to field labour and domestic service much as enslaved women had been, they were not given the factory jobs or the skilled artisan positions which were deemed suitable only for men. The few surviving plantation records indicate that indentured women had a narrower range of tasks on the sugar estates and banana plantations; and they were subject to discrimination in wages. They came to Jamaica during the operation of a system where men began to be paid more than women in spite of the experience during slavery that women survived the plantation experience better than their male counterparts. The contracts signed in the nineteenth century indicate that women were paid 9d.[pence] for a nine hour day and men Is.[shilling] for the same number of hours, though not always for the same types of tasks (C.G.F. 1845-1916: IB/9/3). But the acceptance of a differential rate of pay seemed to have been part of the patriarchal thinking of the period; for the wage differential was made an integral part of the indenture contract even before any tasks were allocated. In any event, the existence of a wage differential was predicated on the assumption that women's work was not as valuable as men's. In 1909, the Protector of Immigrants,

Charles Doorly, informed Governor Sydney Olivier that 'during the first three months of their residence in Jamaica, immigrants are paid a daily wage of men 1/- and women 9d. (a day of 9 hours); 2/6d per week deducted for rations in the first three months' (C.G.F. 1845-1916: 4/60/10A/29, 1909). At the end of three months, theoretically, immigrants could ask to go on task work at rates of pay approved by the Protector; but in any event, it was stipulated that the rates for task work should allow immigrants to earn at least the minimum rates of 1s. for men and 9d. for women. In many cases, the tasks given to female workers were less remunerative than tasks given to men. Weeding, a low-paying task, was traditionally given to women, and many Indian plantation labourers were put in the weeding gang. The only exception was 'heading bananas', which paid 4—5s. per 100 bunches to both men and women. It is not clear from the sources whether men carried fewer or more bunches of bananas on their heads from the fields to the railway siding or the wharf.

On banana plantations, the most remunerative tasks, apart from 'heading bananas', recorded in work allocation books were: forking, trenching, ploughing, lining, circling and cutting. 'Trenching' paid 2—3s. per day and 'forking' 2s. an acre; but not all of these tasks were made available to women. Some men could earn up to 10s. per week from some of these tasks. Picking cocoa, a job that females did, paid 2d. for every 100 pods picked (Sanderson Commission: 1910). On sugar estates, as long as African-Jamaicans were available, they were given the more remunerative tasks. Less remunerative tasks were given to Indian men and the least remunerative to Indian women.

Even when task work was chosen over day labour, female immigrants failed to increase their wages significantly. On some properties, women even earned less than the stipulated rate. Lal's and McNeil's report contains quantitative data which will serve to illustrate this point. These data show that wages received by Indian plantation women in St. Mary ranged from 4¾d. to 9½d., though the upper level was hardly observed. On the thirty-six St. Mary estates in the survey, only two paid the women between 9d. and 9½d. per day or per task. These wage rates represented roughly one-half to two-thirds of the wages of indentured men (Lal and McNeil 1915; Shepherd 1994: 141).

A report on wage rates in 1919 showed that women earned an average of 6s. 11¼d. per week while men earned an average of 9s.

10½d. In 1920, men earned an average of 12s. and women 8s. 6d. per week. The Protector of Immigrants, from time to time, identified outstanding immigrants who earned above this average. Three women—Dulri, Inderi and Jaipali—all earned above 12s. per week in 1920; but in every case, the wages of the outstanding male workers identified exceeded those of the outstanding women - between 16s. and 18s. per week (see JAR 1920). In addition to earning lower wages, the records show that, but do not explain why, female workers were faced with a higher level of expenses than their male counterparts. At a conservative estimate based on rough statistics supplied by Chimman Lal and James McNeil, it would seem that the annual expenditure for females was 76 per cent of annual wages compared with 57 per cent for males (Shepherd 1993: 246).

Women with young children experienced further problems which affected the number of hours they worked and the wages they received. Where neither nurses nor creches were provided, indentured women often had to carry their infant children to the fields. This handicapped them in their jobs and could affect their productivity and therefore the amount of wages they earned. This was the complaint of women on some of the estates visited by the Acting Protector of Immigrants in 1913. He stated that 'recently when I visited a certain estate the indentured women complained to me that it was impossible for them to do a good day's work if they had to take their children to the fields and look after them there' (C.O.R. 1847-1917b: 571/1). A nurse was employed to look after the children and relieve the mothers of childcare responsibilities during working hours after the Protector appealed to the manager on behalf of the women.

On another estate where similar complaints were made by the female workers, the manager agreed to build a creche and employ a nurse to care for the children of immigrant women while they worked. The Acting Protector expressed his wish that, 'all employers of a large number of indentured immigrants ought to be willing to do something of the kind as a great deal more of the time of the women who have children would be available for work' (ibid). But not many estates adopted this practice, arguing that it was too much of an added expense for proprietors.

The result by the end of indenture was a general picture of 'persistent poverty' painted by those who visited the island

and observed the conditions of Indian plantation workers. Such conditions were, arguably, initially worse than those of Indians elsewhere in the Caribbean. For example, J.D. Tyson (1939), an officer on deputation with the Moyne Commission which investigated conditions in the region after the 1930s riots observed that,

> the Indians in Jamaica struck me as the most backward, depressed and helpless of the Indian communities I saw in the West Indies. Their poverty is illustrated by the almost complete absence of property of any kind in their barracks and has the further unfortunate effect that their children are undernourished and are kept away from school for lack of suitable clothing to wear there.

Tyson was particularly concerned about the status of Indian women who, compared to their male counterparts, 'seemed to go for weeks without a single day's work', and whose welfare seemed not to have attracted much attention.

WOMEN, GENDER AND THE DEBATE OVER THE CONTINUATION OF INDENTURESHIP

It was only when the continuation of the system of indentureship seemed threatened by rising costs of importation and repatriation in the early twentieth century that some improvements in the conditions of female indentured servants were suggested. Concerns over a possible labour shortage if labour migration ceased also caused the early twentieth-century immigration rhetoric to reflect a greater pro-natalist stance. Thus, just as the situation of enslaved women featured prominently in the emancipatory rhetoric of the 1820s and 1830s, and just as the improvement of their conditions was enshrined in the amelioration proposals to stem the tide of anti-slavery resistance as well as improve their fertility rate, so gender considerations were critical in the debate over the system of Indian labour which was to replace indentureship.

The discussions over the system of labour to replace Indian indentureship surfaced in the years after the First World War. It was suggested that the emigrants' agreement should be in the form of a civil contract rather than an indenture contract and that the term of contract should be reduced to three years. But the conditions of servitude for females were put at the centre of the debate. Suggestions were now made for women labourers with three children under five years to be exempted from work, subject to the approval of the Protector of Immigrants. It was also proposed that:

> any woman labourer may receive an exemption from work for any particular period either by agreement between the employer and the woman and subject to the approval of the Protector of Immigrants or on the Certificate of the Immigration Department. During advanced pregnancy and after childbirth, a woman may be exempted from work for a period not exceeding six months. Immediate steps should be taken to require the issue of free rations to pregnant and nursing women for a period not less than six months. (C.O.R. 1847-1917: 576/1,1916)

The inducements to be held out to male labourers, though, were linked to efforts to improve their economic welfare. It was suggested that any new scheme of Indian labour after the First World War should include provisions to make land available to male labourers. The recommendation was that all possible steps should be taken to require employers to provide small garden plots of one-tenth of an acre of land for each male labourer and facilities for labourers keeping cows. A larger acreage—one-third of an acre—should be given to those male labourers who were more industrious. This land was to be given after the first six months of labour in the island (ibid). No such considerations were given to Indian females, who in fact were to be encouraged to focus more on family—their 'proper role in life'—rather than on work outside of the home.

THE CONDITIONS OF POST-MIGRANT INDIAN WOMEN

Choice, family ties, the severing of links with India after emigration and tardy repatriation arrangements, all combined to cause the majority of Indians to remain in Jamaica as settlers, the majority continuing as plantation workers. This was despite the fact that on the expiration of their contract, Indians were free to move out of low-paying plantation labour and seek more remunerative occupations.

After ten years in the island, they could also access free or assisted passages to India (once they had not accepted the commutation bounty), or (despite the obstacles) seek higher-paying jobs in Cuba and Central America as thousands of African-Jamaicans had been doing since the 1880s. In reality, illiteracy and lack of access to land kept them tied to rural plantations as agricultural labourers. Only a minority were able to become subsistence farmers on their own plots of land. Some Indian men and women had been helped by the receipt of plots of 10 acres of marginal land in lieu of repatriation; but cash and land commutation grants had been discontinued by 1914. In the grants made in 1903/4, forty-eight women had received land, 67 per cent of them being single women. In the 1906 allocation, sixty-nine men and thirty-eight women got land grants (Shepherd 1994: 121—3). The women who came to the island after 1910 would have had to buy land on their own account or cultivate family land; and this was not only expensive but access was tightly controlled by the planter-class. Continued labour on rural agricultural estates was the only option for many Indian men and women, only a-minority making it in the non-agricultural arena up to 1943.

In 1891 when the Indian female population numbered 4,467,58.4 per cent were agricultural labourers. A further 399 or 9 per cent were general labourers, 151 were shopkeepers and 113 were doing household/domestic jobs. Only a few were numbered among the professional, industrial and commercial sectors which in any case, were male-dominated. For example, there was only one Indian female teacher. Appendix 3 illustrates the nineteenth-century occupational pattern.

According to the 1911 as well as previous censuses, the majority of Indian females worked as agricultural labourers mostly in the banana and sugar industries (see Appendix 4). On the all-island level, there were 49,116 females in agricultural labour, including 3,734 out of 7,452 Indian women. A significant proportion worked as domestic servants. On the all-island level, there were 35,701 domestic servants of which women made up 30,316. Of 188 Indian domestic servants, 134 were women; an additional ten Indian women worked in other domestic or personal service.

In the 1921 census, 3,828 Indian women were engaged in agriculture, again, the majority on plantations. Only one Indian woman was represented among the 144 rangers and supervisors on agricultural properties in the island. Among the peasant farmers, male Indians dominated. Of 188 Indian banana farmers returned in the 1921 census, there were only thirty-one females. There were thirteen females out of sixty-two Indian cane farmers; two out of thirteen cocoa farmers; 107 out of 399 provision farmers; forty out of 110 rice farmers, and seven out of thirty tobacco farmers (even though more females than males worked on tobacco farms).

Employment opportunities for women increased only marginally as a result of male emigration from the 1880s. The migration wave was dominated by African-Jamaican men and the gap created by their emigration was increasingly filled by Indian men, with African-Jamaican and Indian women getting work where male labour was not available. Even so, any such new employment opportunities were available mostly in agriculture.

By 1943, as revealed by the official records (1943 census) 56 per cent of Indian women were agricultural labourers, most still confined within the latifundial environment of the plantation. In contrast, only 28 per cent of African-Jamaican women, 1 per cent Chinese and 1 per cent Syrian women were returned as agricultural labourers on plantations. By contrast, whereas 56 per cent of Syrian women and 49 per cent of Chinese women were in the retail trade, only 12 per cent of Indian women were similarly engaged.

When one considers the general absence of schooling among Indians, particularly among the girls, the occupational pattern becomes more understandable; for there seemed to have been a close correlation between the educational level of the wage-earning population and the industry to which they were attached. Of 2,145 school age Indian children in 1891, only 110 girls and 126 boys were attending school (see JCR 1871-1943). A combination of social and economic factors militated against a more significant school attendance among both African-Jamaican and Indian-Jamaican children. By 1943, 49 per cent of the Indian population was returned in the census as illiterate. This was in contrast to 28 per cent African-Jamaicans, 14 per cent 'coloured', 3 per cent whites, 14 per cent Chinese and 6 per cent Syrian (see JCR 1871-1943; Shepherd 1993).

The economic crisis of the inter-war and post-war years also had an impact on the conditions of agricultural workers. The world-wide economic depression, the cessation of emigration as an outlet for those seeking higher wages and better living conditions, the return of migrants to swell the ranks of the unemployed, the termination of government-sponsored repatriation of Indians, failure of the planter class to heed the workers' call for higher wages and better local working conditions had all culminated in widespread unrest in Jamaica and the wider Caribbean in the 1930s (Post 1978; Shepherd 1994: 142). When J.D Tyson visited the island in 1939, he reported that there was a scarcity of work among women who formerly found jobs on coconut, banana and sugar estates. Many women went for days without work at a time when close to 60 per cent of them depended on the plantation for employment. According to Tyson (1939: 6), 'the complaint of short work for Indians - all the time on the banana estates and out of "crop" season on the sugar estates—was general wherever I went. Indians on banana and coconut plantations were thus living in a state bordering on destitution. Those who got 2—3 days' work per week and made thereby from 3s. to 5s. were considered fortunate.' Indian women were discriminated against in the allocation of such scarce jobs. Tyson (1939: 33) explained why this was so. He admitted that Indians suffered along with the rest of the labouring population of the island from the general wave of unemployment; but that, in the case of the Indians, this had been made worse by 'a growing competition in fields hitherto regarded as his own, from West Indian labour returning from Cuba and elsewhere with some training and experience in estate work'. In addition, there were allegations that headmen on the estates who were generally African-Jamaican, tended to 'favour their own people in the distribution of piece work, especially where there was not enough to go around'. He added that 'unemployment is general in the island but Indian labour has been especially hard hit by limitation of production in the two crops in which this labour is principally utilized—sugar (owing to the quota) and bananas for which, owing to the prevalence of various banana diseases, coconuts have been substituted'. The substitution of coconut for banana cultivation reduced the demand for labour and deprived the labourers of the customary use of marginal estate lands for their own cultivation of subsistence crops. Indian labour was, therefore, displaced and lacked the fluidity to seek alternative jobs in non-agricultural fields. Some found outlets nevertheless in rural-urban migration. Predictably, the parishes experiencing the highest drain were the depressed agricultural parishes of

St. Thomas, Portland and St. Mary—also the most contiguous to Kingston and St. Andrew. In 1943 when Indians numbered 21393, representing 2.1 per cent of the total Jamaican population, 8 per cent resided in Kingston and 14 per cent in St Andrew. They were still very much a rurally settled population, with 76 per cent depending on wage labour; and they were only 2.6 per cent of the total urban population. The majority of those living in Kingston and St. Andrew were female who sought economic opportunities in the cultivation and door-to-door sale of flowers, fruits and vegetables, or worked as domestic servants in the households of the urban elite. Indian women were 2,297 or 52 per cent of all Indians settled in St. Andrew in 1943. By contrast, 33 percent of Chinese women lived in Kingston and Port Royal and 18 per cent in St Andrew (Shepherd 1986 and 1994: 135).

CONCLUSION

In conclusion, I wish to admit that the task of uncovering the historical experiences of Indian plantation women is not an easy one; for the colonialist historiography has often been 'gender neutral', keeping the working class, the subaltern, 'mute'. Indian diaspora women's experiences have been constructed less by themselves and more by those who claim to speak for them. The sources are predominantly official; and the 1000 letters plus, from and about Indians, which are stored as part of the Protector of Immigrants' files in the Jamaica Archives, though representing a potentially rich source for listening to the voice of Indian women, were mostly written by others on behalf of the immigrants and settlers. Furthermore, from the perspective of this topic, of the 1,858 letters selected for analysis, 77% were written by, to, or on behalf of men; 14% concerned female immigrants and settlers specifically. The concerns of Indians living in Kingston and St Andrew predominated, accounting for 67% of the letters compared to 25% relating to Indians in rural Jamaica, not all relating to plantation labour.

From the records available though, it would seem that the Indian female experience of migration and indentureship was conditioned by the colonisers' own perceptions and prejudices, and by race and class factors; but it is undeniable that gender discrimination was a visible element. The landholders' preference for male labourers, and their irrational belief in the inefficiency of female labourers, led to a gender-specific immigration policy; and the sex-typing of jobs and the accompanying wage differentials all helped to shape the Indian female experience in Jamaica. The types of occupations of female Indians in the period after the end of the system of indentureship reflected the lack of an educational standard necessary to equip them for higher-status jobs. It also reflected the ideology of the day which sought to confine women to 'female type jobs'. Up to the 1940s, the majority of Indian women were still tied to rural estates as plantation labourers, receiving lower wages than their male counterparts. This weak position of women in the labour market encouraged by wage differentials and the promotion of the male breadwinner ideology — the same notion of the 'proper gender order' which missionaries tried so hard to instill into African-Jamaican women in the immediate post-slavery period-was still in place up to 1943. After years of working as plantation labourers, Indian women were less able than other immigrant women to experience upward social mobility. Yet, despite their socio-economic marginalisation, after 1943, some did manage to break free of the plantation system, using educational opportunities and commercial enterprises to seek economic autonomy. As plantation workers they also used strategies of resistance to subvert the attempts of the landholding class to subject them to the exploitative elements of the indentureship system.

APPENDIX 1.

The Indian population in Jamaica: Male/Female 1871-1921.

Census Year	Male	Female	Total	Per Cent of females
1871	5,339	2,454	7,793	31.5
1881	6,941	4,075	11,016	37.0
1891	6,338	4,467	10,805	41.3
1911	9,928	7,452	17,380	43.0
1921	10,203	8,407	18,610	45.2

Source: Jamaica Census, 1871-1921.

APPENDIX 2.

Distribution of Indian Females by Parish: Selected Years

Parish	1881	1911	1921	1960
Kingston & Port Royal	54	196	357	446
'St. Andrew	196	219	292	3,018
St. Thomas	206	692	626	624
Portland	189	610	627	630
St. Catherine	824	1,145	1,439	2,056
St. Mary	659	1,865	1,856	1,652
St. Ann	21	11	25	68
Clarendon	732	1,070	1,215	2,313
Manchester	34	36	30	88
St. Elizabeth	139	77	102	158
Westmorland	779	1,275	1,461	2,649
Hanover	111	108	171	126
St. James	59	79	147	163
Trelawny	72	69	59	35
TOTAL	4,075	7,452	8,467	14,026

Source: Jamaica Census, 1881-1921.

APPENDIX 3.

Occupations of Indians in Jamaica: Male/Female, 1891

	Men	Women	Total
Peasant proprietors	59	3	62
Overseers	15	0	15
Agricultural labourers	3,707	2,534	6,241
Attending stock on pasture	10	0	10
General labourers	523	399	922
Attending agricultural machinery or boilers	1	0	1
Merchants/Agents/Dealers	10	0	10
Shopkeepers	246	151	415

	Men	Women	Total
Shop/Sales/Clerk Market	39	46	85
Gardeners Indoor	60	50	110
Domestic Servant	35	50	74
Washers	0	26	26
Interpreters/Messengers	4	0	4

Source: Jamaica Census, 1891.

APPENDIX 4.

Summary of Occupations of Indians in Jamaica: Male/Female, 1911

	Male	Female	Main Categories
Professional	35	11	Students, Nurses, Teachers
Domestics	91	152	Indoor house servants
Commercial	386	204	Barkeepers, Peddlers Shopkeepers, Store Servers
Agricultural	6,649	3,734	Wage Earners on Plantations
Industrial	165	78	Skilled Trades, e.g. milliners, washers
Indefinite and unproductive*	2,602	3,273	

Source: Jamaica Census, 1911. * Included women working at home!

APPENDIX 5.

Percentage Distribution of Indian, Black, Coloured and Chinese Men by Parish, 1943

Parish	Indian	Coloured	Chinese	Black
Kingston and Port Royal	7.7	18.1	32.9	7.4
St. Andrew	13.8	14.7	17.8	9.3
St. Ann	0.4	6.7	3.3	8.2
St. Catherine	13.8	5.8	6.9	10.5

Parish	Indian	Coloured	Chinese	Black
St. Elizabeth	1.2	11.6	2.5	7.5
St. Mary	17.6	5.2	5.0	7.4
St. Thomas	5.8	1.9	3.7	5.4
Clarendon	13.2	7.2	6.9	10.4
Hanover	1.9	4.7	1.7	4.1
Manchester	0.7	6.8	4.3	7.9
Portland	5.6	3.3	5.1	5.2
Trelawny	0.7	3.3	2.4	4.0
Westmorland	16.4	7.1	3.0	7.1
St. James	1.2	3.9	4.5	5.6
Total Per Cent	100.0	100.0	100.0	100.0

CHAPTER SEVEN

The Depositions

LISA YUN

A Chinese glowered like a spark of fire amid gray ashes; his usual expression of sullen insubordination being sharpened by the pressure of physical suffering. One of these sat on the edge of his bed, with, a swollen and bandaged limb drawn up beside him—the very incarnation of impotent hate and rage. The mayoral laid a firm, detaining grasp on his shoulder, under which I could see the man wince and shiver, while the official told me how he had run away weeks ago, and hidden in the woods, leading a sort of highwayman's life, and baffling all pursuit, until he cut his foot badly on a sharp stone, in jumping a stream; which wound festered and gangrened, and so disabled him that he could no longer procure food, nor drag his wasted body from one hiding place to another; when he was found—half-dead, but untamed in spirit—and brought back to prison. Since which time, he had twice attempted suicide. The Chinese meanwhile regarded us with a look that would have stabbed us both to the heart, if looks were available for such a purpose. Plainly, he felt himself at war with the whole tyrannous universe; and especially resented the indignity of being exhibited and commented upon as if he had been a wild beast.

—Julia Louisa M. Woodruff
(pseudonym "W.M.L. Jay")

In 1871 Julia Louisa M. Woodruff published her observations of her time in Cuba. Included was her visit to the Santa Sofia sugar plantation, where she encountered a recaptured coolie. Her description touched upon the power struggle inherent in domination and daily resistance. "Half-dead and untamed in spirit," the coolie represented a continual challenge to management and its drive to maximize output through captive labor. While coolies appeared in observations of white travelers and journalists of the time, what did the coolies themselves say about these daily struggles? On the whole, the depositions did not usually present the sort of lengthy arguments presented in the petitions. However, they provided detailed accounts of the coolies' daily trials and conditions, and revealed startling views of how systemic subjugation was designed yet resisted. Some coolie deponents expressed intense resentment and described the planned murder of their overseers. Others relived their humiliation, such as one who described being forced by the manager to bark like a dog and bleat like a sheep. And there was another who numbingly declared, "I just cannot stop crying" (Deposition 938). Those such as Wu Axiang plaintively yearned to return to their families: "I really want to go back to China and see my mother" (Deposition 114). Occasionally, others revealed moments of youthful naïveté. One Californian named Yang Atian, age twenty-six, was rueful about his situation. He had lived in San Francisco since he was fourteen years old. After working on the railroad in California, he took work on a ship, with a misguided sense of adventure. To his dismay, he lost his earnings by gambling at the port of Havana and was forced into bondage. He concluded, "My four brothers and my parents are still alive, but they do not know I am here. I really walked right into the trap myself!" (Deposition 1036) The texture of individual experiences ranged from this naïve youth to the sixty-six year old Yuan Aishan who recounted his escape attempt and his capture by soldiers (Deposition 378). The rawness and spontaneity of the depositions contrast with the deliberate, collective writing of the petitions.

Of the 2,841 testifiers, 1,176 gave oral testimonies, and some explicitly named their masters or their plantations, such as the plantations of "San Antonio," "Recreo," and "Esperanza" (in Cardenas); "Juniata" and "Candelaria" (in Cienfuegos); "Espana" and "Flor de Cuba" (in Colon); "Santa Ca-talina" and "Las Canas" (in Havana); "Armonia," "Concepcion," and "San Cayetano" (in Matanzas); "Santa Ana," "Capitolis," and "Santa Isabella" (in Sagua la Grande). Each of the oral testimonies were unique yet bore out shared themes. One common thread concerned the micro view of daily survival, portrayed as tenuous and unpredictable. The acute conditions of deprivation and the nature of resistance within those conditions were described fey the coolies in the context of *expendability* and *racialized conflict*. How coolies resisted their forced bondage, in the midst of shifting political conditions and uneven relationships to power, can in some ways be compared to forms of resistance that emerged 'in histories of oppression and labor. Coolie resistance manifested itself in passive recalcitrance and covert sabotage, as well as overt individual and group rebellion. Explications and arguments on what constitutes (or what can be read as) "resistance" especially appear in treatments of resistance of peasantry, contract labor, and slavery. These treatments help highlight the particularities of coolie resistance in Cuba. James Scott's foregrounding of covert forms of peasant resistance, "weapons of the weak," has expanded what were conventional narrations of resistance as open, collective, or large-scale revolts. By attending to the local specificities that shape the overriding concerns of the oppressed, Scott raises questions regarding how we define resistance and how local conditions make those forms particular. Howard McGary also undermines conventional definitions of resistance through a philosophical critique of the emphasis upon "intent" as the measure of resistance. He notes, "We often associate acts of resistance with courage, but it would be a mistake to claim that every act of resistance is an act of courage." As a case in point, he examines slaves stealing from their masters as a form of resistance. Scott's and McGary's conceptions of resistance extend to subversive actions beyond overt acts of rebellion and beyond acts deemed "courageous." In a history of plantation labor of Hawaii, Ronald Takaki also drew attention to forms of resistance manifested in daily acts. Workers organized strikes, but according to Takaki, they also employed forms of daily resistance, including intentional laziness, inefficiency, and work slowdowns. Scott, McGary and Takaki assess the different histories of peasantry, slavery, and contract labor, and present "resistance" as a range of actions. The critical interventions on resistance mitigate what George Rawick. Once called the "entire view of the slave as Victim and Object."

However, these analyses assume certain continuities and foundations. The peasantry in Scott's study is grounded to the land, and wedded economically and socially to the village. McGary's critique refers to the "peculiar institution," an institution characterized by generations of slavery and an equally long history of slave resistance. Contract labor in Takaki's study was relatively rooted, in comparison to contract labor in Cuba, and was shaped by planters' designs to form a stable (and hypothetically less volatile) labor pool. The extensive presence of contract labor families on the Hawaiian plantations led to social and cultural patterns of continuous plantation and migrant life that markedly differentiate the Hawaiian history of contract labor from the Cuban history. The constellation of examples and arguments regarding resistance exposes a problematic lacuna for grappling with representations of resistance in the coolie testimonies of Cuba, in which resistance was presented under the chief conditions of high attrition, disposability, and hypermobility. As one Liu Arui declared, "The owner often said to the overseer, 'I only care about how much sugar they (Chinese) can make, and you don't have to treat them well. If you beat one Chinese to death, I still can afford to buy ten more next year.'" (Deposition 244) Coolies were mobile slaves with short life spans, and mobility appears as a main trope of their testimonies. Contrary to liberal narratives of mobility (which emphasize physical and social mobility as basic features of free society), the combination of mobility with disposability emerged as a form of enslavement. Narrations of resistance in the depositions were shaped by this condition of mobile enslavement and, furthermore, shaped by the racial hierarchies that structured this very condition.

RACE, RESISTANCE, AND SPECTACULAR SUBORDINATION

Overt resistance, including heightened and deadly conflict, emerges as a distinct and pervasive feature of the testimonies. The divisions between yellow, black, and white, and survival under extreme exploitation, were revealed in the testimonies as everyday struggles that often led to deadly consequences. For the treatments of Asian and African histories, this aspect reveals a unique dimension to histories of resistance, racial conflicts, and collaborations. Here, more attention is paid to overt resistance in the testimonies, though some attention is paid to representations of passive and covert resistance. The testimony of twenty-three-year-old Ou Rong, abducted and shipped to Cuba at age fifteen, offers, an example of mixed forms of resistance deployed by the coolies, such as running away, altercations with owners, refusal to work, and taking sick:

In the fifth year of Tongzhi (1866), someone invited me to see a performance in Macao, but I was taken to Xinhe Pigpen. I lived there for two weeks and met with a foreign officer, signed a contract and was given eight dollars. The ship set off in the ninth moon (October, 1866). Two months after arriving at Havana, I was sold to a family in the city as a waiter. I worked there for nine months. Later I had a quarrel with the son of the owner, so I was locked up while working for three weeks and got fiercely beaten. Then I was sold to a sugar plantation where I was treated badly. I did not want to work there, but the manager told me if I did not want to work, I would be locked and put in chains. I said, "I am not afraid of wearing chains. I am not afraid of you stabbing me with a knife. I don't want to stay here." So he sold me to this sugar refinery in Cardenas. This sugar refinery treated me badly too. The food was not enough and the work was toilsome and difficult to do. In the sugar refinery, I knew eight people who hanged themselves. I got sick a few months ago. I told the overseer I was sick, but he just beat me. Then I ran away. I stayed outside for three weeks but then was captured. I have been chained for a long time since they caught me. (Deposition 776)

Within seven years, Ou was sold into a family household, then to plantation labor, and later to a refinery. With each type and location of bondage, Ou Rong resisted in different ways, including his striking description of speaking out. His recounting of passive and overt disobedience demonstrates various forms of resistance on Ou's part; but his testimony also brought to light the continued problems faced by coolie owners in their attempts to control a new force of indentured labor. Ultimately, Ou resisted by running away, though fugitive coolies like Ou were often recaptured by *rancheros* and dogs, or by patrols and soldiers. The only manner of temporarily controlling Ou appears to have been chaining. Unlike Ou however, those who succeeded in escaping were not represented in the testimonies and are not represented in marronage studies, thus leaving a gap in knowledge regarding how Chinese might possibly have formed, joined, or collaborated in maroon communities. (However, the testimonies do include references to attempts to form organizations of coolies across plantations and covert resistance in the form of alliances with sympathetic "foreign" officials, as discussed in the previous chapter on the petitions).

Besides his own resistance, Ou Rong also mentioned having known "eight people who hanged themselves." The accounting of suicides is a prominent feature of the testimonies. Given its preponderance, suicide had undeniable effect upon those who *witnessed* these acts. Like the witness petitions, suicide accounts conveyed a desire to bear witness for those who did not survive and revealed conditions that contributed to overtly violent resistance. He Aying succinctly listed the types of suicide, which were commonly described throughout the testimonies: "There were people who hanged themselves or cut their throats; there were people who took poison or drowned themselves in the river" (Petition 73). Chen Ming observed those who dashed their heads against a well: "People cut their throats, jumped into the river, or smashed their heads to the well; some who wore shackles and suffered a wrong took poison" (Petition 9). Li Wencai even detailed how dead persons were retrieved: "I witnessed people hanged themselves in the woods, and people who jumped into the well were taken out of the well by an iron hook" (Deposition 116). It is tempting to read such accounts of suicide as narrations of despair in a history of exploitation. As summed up in a petition by Wang Hua and companions: "The pain and sorrow are so unbearable that we have no release until death" (Petition 42). Or perhaps, this pattern could be read as a cultural (Chinese) predisposition toward suicide, a misguided analysis at best. The South African paper *Transvaal Weekly Illustrated* once featured such a presumption, in regard to the local suicide of a coolie who protested the discriminatory Asiatic Law Amendment: "The Chinese ethics of suicide are certainly incomprehensible to the European."[6] However, Leung Quinn, leader of the Chinese passive resistance movement at Transvaal and a colleague of Gandhi, responded that he "was not aware of suicide being common in China. The Chinese regarded human life as valuable and only ended it when they are driven to desperation by much the same causes as induce suicides amongst Europeans."

Positing a cultural predisposition for suicidal depression is unsatisfactory, most obviously in regard to testimonies that suggest suicide as collective protest and workplace sabotage. These accounts included the witnessing of suicide as group acts. A nineteen-year-old student when he was abducted and shipped to Cuba, Wen Changtai recounted as follows:

I *witnessed nine workers who hanged themselves, one who jumped into a hot sugar cauldron, twelve who died because they were flogged so hard that their flesh festered and had maggots, and some who ran into the mountains and starved to death. I do not even know whether some people are still alive or not after they escaped from the sugar plantation. I saw lots of human bones while working on the sugarcane fields. I witnessed some workers who were bitten to death by the manager's dogs.... Once I saw seven people kill themselves by jumping into a well. Countless people hanged themselves or were beaten to death. White people here treat Chinese worse than dogs." (Deposition 18)*

The testimony of group suicide (seven people jumping into a well) potentially raises the macabre question as to how this act was planned and executed. Yet a history of the Chinese in Cuba mentions a similarly disturbing incident of fourteen coolies who committed group suicide on a plantation called "Dos Marias."[8] Henry Auchincloss noted another incident, this one concerning the predicament of a coolie owner in Cuba who sought measures against coolie suicides. The owner found that his "coolies were destroying themselves at the rate of two per day or more" and that one "eccentric Celestial climbed atop the highest chimney on the hacienda and hung himself in full view of the entire body of his colleagues." The owner responded to this public display with a spectacle of his own, by burning the coolie body "in full view of the entire plantation" and scattering the ashes "to the winds."[9] This incident suggests the coolie body as also a body of public spectacle, not only as colonized labor but also as a culturally and racially signified body. The struggle for power here depends on cultural subjugation, as the owner did not mete out physical punishment; instead, he sought to deeply violate Chinese cultural beliefs, which do not espouse the scattering of body and spirit.

That some suicides were committed as collective action and as public display suggests that a dimension of protest, and not simply a predilection for despair, was inherent in the patterns of coolie suicide. At certain times, coolies declared they would rather die than submit to the outrage: "Several workers who had been constantly beaten hanged themselves because they could not bear the humiliation and rage" (Deposition 58). One of the most frequently mentioned methods of suicide could be found in the workplace, where coolies would sometimes jump into the "hot sugar pot" — a sugar cauldron kept at extremely high temperatures to boil and purify the sugar. Chen Guanzhi's testimony typified the accounts of "sugar pot" suicide: "I witnessed lots of people who committed suicide by jumping into the sugar pot, jumping into the refinery, machine, hanging themselves and jumping into the well.... Also I saw two people chained together, for they tried to run away. They still needed to work, and the manager hit them severely. Then these two people couldn't take it anymore so they jumped into the sugar pot together. I witnessed that as well" (Deposition 816). The sugar purification process, in which a hot cauldron was crucial, was a worksite process that coolies and slaves manned in the sugar house or sugar mill, in addition to their arduous labor in the cane fields. In *el ingenio*, there would have been a series of hot pots or a sugar train. Richard Henry Dana once described this process in the following manner: "From the last defecator, the juice is passed through the trough into the first cauldron. Of the cauldrons, there is a series, or, as they call it, a train, through all which the juice must go. Each cauldron is a large, deep copper vat, heated very hot, in which the juice seethes and boils."[10] Suicide by jumping into a seething boiling cauldron would have disrupted the immediate sugar processing and could be read as a gruesome act of protest and workplace sabotage, as it had an undeniable effect upon witnesses and disrupted the worksite.

In some cases, other forms of workplace sabotage and resistance were implied. Testimonies of "mysterious" incidents of fire in the sugar refineries or mills were conspicuous enough to suggest their being veiled references to forms of sabotage and resistance. For example, at the Las Cañas plantation, Chen Yiyou noted that "the plantation caught fire twice before, and everyone else and I tried to put off the fire. But the manager suspected that someone set the fire, so he investigated for a long time but he could not find any evidence. How the fire was started has remained mysterious until now" (Deposition 938). Others mentioned similar fire incidents caused by unknown parties. Another coolie stated that "the wood floor in the sugar plantation caught fire. The manager said the fire was set by a man from Xiangshan. He denied it" (Deposition 729). One coolie emphatically denied his involvement in a fire in the following manner: "Once I was beaten extremely hard because the overseer thought I set a fire in the sugar plantation. But I am truly innocent" (Deposition 197). Chen Dezheng also asserted his innocence; yet called attention to such conflagrations when he stated that "the plantation got fired twice. I was shocked and woke up to help others put off the fire. I don't know what caused the fire" (Deposition 439). Some testifiers even described worksite killings as "accidental," As in the case of Li Hui who claimed that he unintentionally killed his overseer: "The overseer hit me with a handspike. I took that handspike and pushed him; however, the overseer fell into the sugar-cooking pot. I did not mean to kill him." (Deposition 67)

Alongside "accidents" and acts of sabotage, overt confrontation emerges as one of the most noticeable patterns in the depositions. Testifiers indicated that their propensity to act, often collectively, was prompted by witnessing suicides and deaths. They were moved by righteousness but also by self-preservation. As one coolie put it, his group killed the overseer because "we saw many people were beaten to death on the sugar plantation, so eleven of us talked and concluded that if we did not kill the overseer, we would end up being beaten to death by him" (Deposition 503). Chen Asan came to a similar conclusion: "We'd rather risk our lives killing him than be killed by him. So. we beat him with hoes. The white overseer was injured and died after seven days" (Deposition 715). In other cases, young, healthy Chinese laborers were incensed by the abuse not only directed toward themselves, but also toward those who appeared weak, handicapped, or old. Zhu Afu described the accumulating outrage, intensified by the suicide of a fellow coolie who was brutalized by an overseer: "Three workers and I were so angry that we beat the overseer with hoes. He died that night in the ward" (Deposition 559). After eight years of labor, Liang Agui exploded under the hot sun upon seeing the manager beat two people: "At'my eighth year, in May it was so hot, but I still did not dare to stop working. But the manager still said we didn't do our jobs well and beat two people severely. I was so angry. I beat him to death with my hoe" (Deposition 880). Similarly, the case of "Hung Aguang" is a good example of how Chinese workers witnessed the abuse and suicide of a fellow coolie and were then motivated to collectively seek justice:

> Hung Aguang couldn't stand the torture and ran away, but he was caught. He was locked up and beaten so hard that he bled all over. He still had to wear chains at work... They barely gave him any food. He was so hungry that he ate sugarcane on the

sugarcane field, and when the overseer spotted him eating the
sugarcane, he was beaten again immediately. Hung Aguang
hanged himself.... Then more than twenty workers went to the
official and complained; they said they did not want to work in
that sugar plantation anymore. (Deposition 670)

Comparable to Huang Shirong's account is Lai Axi's account (Deposition 659). Lai elaborated on the same incident but placed greater emphasis on protest. He emphasized that the Chinese would go to one official who did nothing and then went to a second official, with the manager following close behind. In this instance, twenty coolies protested plantation conditions, although in the end, this obviously did not change their circumstances. Nevertheless, Huang and Lai pointed to this protest as a defining moment that motivated them to openly seek legal redress, despite legal routes being overwhelmingly depicted by the coolies as corrupted. The words of former sailor Chen Amu typify the accounts of resistance through legal routes, with outcomes that apparently worsened their situations:

"On the sugar plantation, I have been treated atrociously. I am hit all the time. Last year I was hit so hard that I could not take it, so I went to the local officer and sued the owner. But the owner bribed the officer with money and reclaimed me back. I was shackled for months, and I was treated more horribly than ever" (Deposition 905).

The propensity to appeal to legal process may have stemmed from the cultural tradition of legal institutions in China; but the coolies were also aware of being categorized as "contract" laborers who supposedly had some legal recourse (also see previous chapter).

The awareness of this rhetorical and legal differentiation, as contract labor and not slave labor, was also accompanied by feelings of outrage and humiliation. The Chinese, even those who were not of elite backgrounds, believed their stations to be above that of manual laborers, and certainly that of slaves; They thought of themselves as more like visiting craftsmen. As one' coolie stated, "In China, I have been to shipyards in Shanghai and Fujian, where we treated foreign craftsmen with reason and courtesy. I can't believe that when we Chinese work abroad, we have to suffer torture like this" (Deposition 881). In a related form, humiliation was expressed out of national pride, with Chinese describing their subjugation as a synecdoche of

China's weakness under foreign incursions and imperialisms. This was particularly evident in the common accounts of nakedness and queues being cut off when being sold, such as farmer Chen Aji's description of being hit "because I refused to cut my hair off " and Lin Abang's memory of humiliation upon being sold: "Chinese people who were sold in the Selling People House in Havana were forced to take off all their clothes in order to be examined whether they were healthy and strong or not. They were treated just like bulls and horses. Chinese people here were not only physically abused and tortured, but also mentally humiliated" (Depositions 4 and 74). Further compounding their feelings of humiliation and injustice was the devaluing of Chinese language, as Chen Abao recounted: "Because I could not understand the language, I was beaten a lot" (Deposition 217). Chinese ethnic identity and the rhetorical differentiation for "contract" labor did not translate into social privileges, and according to the Chinese, quite the opposite.

However, humiliation was most pointedly detailed in accounts of subjugation as spectacles, and coolies recalled the ensuing rage that such treatment spawned. "They humiliated us in every possible way," said coolie Wu Asan (Deposition 397). The spectacles of their enslavement encompass what Saidiya Hartman calls the quotidian and "benign" acts of entertainment, as well as the brute violence wrought under the master and white audience. Slave revelry, which Hartman gives as an example, was also entertainment for the master who watched slaves and commanded them to dance and sing. Although the testimonies do not suggest a history of theatricality and performance in the brief generation of coolie labor, accounts similar to Wu's do depict forms of spectacular subordination or theaters of subjugation, such as the ritual of denigration administered to Chinese at the Flor de Cuba plantation: "Every Chinese who was locked up was forced by the manager to bark like dogs and bleat like sheep. If we refused to do so, we would be beaten severely. They humiliated us in every possible way. Sometimes I do not even have clothes to wear. My life here is like a living hell" (Deposition 397). The forced imitation of dogs and sheep served to reduce Chinese to the level of domesticated animals. Similarly, coolies recounted the humiliation of having to plead and beg for various privileges but most of all, for their freedom papers. Though unlike the spectacles of beating and physical domination, the recurrent spectacles of begging extended relations of domination and subjection. Chen Gu stated, "I begged them on my knees to return my freedom paper," although he was hauled off instead to an official workshop (Petition 22). Deng Asi, consigned to prison after being framed for "forgery" of a freedom paper, stated the following: "I begged them to release me many times, but they refused" (Petition 46). Xie Zhiren, unable to write his name in Spanish and dependent upon the will of his master, "begged the master to write down my name and ask for a paper" (Petition 71). In some cases, coolies begged for other coolies, as in the case of one who begged for his friend's release, after his friend attempted to escape a cruel master but was later caught and locked up (Petition 17). Depictions of pleading and begging, however, were as frequent as depictions of confrontational violence. The pleas of coolies alongside patterns of confrontation underscore the different forms of resistance that were in fact related. Coolies "turned"'from begging and pleading to taking sick or, in some cases, to plotting the murders of their overseers, such as described in Liang Lianqing's blunt recounting: "Once two other workers and I were all sick and asked to rest, but the overseer denied our request and still required us to work. Moreover, he beat us severely. So three of us killed him with hoes in the sugarcane field" (Deposition 1010).

The accounts of resistance expose conditions of "contract labor" as both sites of struggle for domination over unwilling subjects and as sites of racialized pleasure and abasement. One coolie quoted the sentiments of his third owner, an especially cruel man, who said, "I did not buy you to work for me; I bought you because I want to beat you" (Deposition 199). Yuan Aan described an overseer who "was so mean that he actually enjoyed beating us" (Deposition 120). Another named Chen Ajin, formerly a ship carpenter, described the torture inflicted upon him at the will of his overseer: "One time, I told the overseer I was sick. He said I was faking it and told four people to truss me up, take off my pants, and beat me two hundred times. My flesh festered. Then I was put in shackles and still had to work. But that was not over. They came back at night and forced salt and orange juice into my wounds. I almost died of the pain" (Deposition 120). Coolie torture and abasement further extended to entertainment at the hands of "little white and black children," who would throw stones at coolies upon seeing them in the streets (Deposition 144). Accounts revealed the features of *subordination as spectacle* and the exercises of power conscripted the least powerful of the social spectrum, that being children and the coerced participation of both

Africans and Chinese. In a ritual of abasement, one owner forced Chinese to drink the urine of female African slaves, as recounted by a coolie named Liang Aren: "Later I was sold to a sugar plantation where I was always beaten. The owner was atrocious. If the [Chinese] worker was sick, the owner would ask a black woman to urinate and then force the worker to drink it. If the worker drank the urine, then he would be considered really sick and could have a rest. Otherwise, he had to keep working." (Deposition 175)

This ritual-as-spectacle turned upon mechanisms that inscribed racial and gendered hierarchy. Liang was subjected to drinking urine, but more specifically, to drinking the urine of a black female slave, which was constructed or *presented* by white management as the lowest or most denigrated form of the body (the act of drinking urine itself, outside this particular historical context, is not necessarily "humiliating"). In this scenario, the male Chinese coolie was coerced into a status "inferior" to that of a black female slave and thereby humiliated—an act which would have been effective if made as an example, or spectacle, for others. A momentary saving face or saving of "manhood" would have been perpetrated through the coolie's refusal of black femaleness. The enactment of this rite reproduced the ideology of white superiority/black inferiority, with the Chinese coerced into participating in a racist schema, thus coercing an entire social body more effectively than an actual whipping. The black female slave, meanwhile, was subjected to humiliation by urinating (for the satisfaction of the owner) in a spectacle that debased both her and the coolie and that simultaneously induced antipathy, disgust, and the prurient desire of the spectator. Rather than whipping the coolie into submission, another form of violence, the spectacle of humiliation, effectively inscribed and reproduced hierarchies of domination and inferiority in the social body, and promoted interracial anxiety and conflict through racially charged and gendered violence.

Daily terror also emerged in theaters of racial, gendered, and sexualized acts, such as the daily terror of being gang-whipped, inflicted upon Chen Afu. Chen, abducted and sold to Cuba at age eleven, described being regularly abused by an overseer who liked to beat him while he was held down by four "others": "I was sold to a sugar plantation to herd horses. The overseer was very mean and hit me a lot, for I was young among most workers here. He would ask four other people to push me down to the ground, take off my pants, and hit my bare bottom with cane" (Deposition 12). Rituals of terror and abasement were described by Chinese as attempts to break them psychologically and physically. Beyond the daily whipping and chaining, these methods took several forms: cultural and sexual subordination in racialized contexts (such as hair being cut off, being beaten for not speaking Spanish, clothes taken off, being held down and having their pants taken off, being beaten for "pleasure," being forced to drink the urine of a black slave, and being told to imitate animals), psychological and physical terrorism (the preponderant use of attack dogs, pouring salt into wounds, pouring boiling water on the body, cutting off ears, chopping off fingers, public torture, stripping off clothes), coercion through imminent disposability (the threat of being arbitrarily beaten to death like the "others," being told that they could be disposed of at any time, and being told that they would be sold into a worse situation), and withholding (of food, water, and most of all, freedom papers).

Cruel punishments were made visible and marked the coolie as a subjugated body. An ear chopped off, feet gnawed by dogs, flesh festering with maggots, fingers chopped off, skin burned—all revealed the bodily visitations of domination meant for long-term display. Cruel acts on the part of the owners such as ordering attack dogs to bite coolies (dogs obviously had to be trained to do such things), terrorizing and gang-whipping a boy in the field, and forcing a man to drink urine, all comprised a litany of the perverse realities beneath contract labor, "transition," and "modernization." Appearing before the commission, coolies offered their bodies as evidence. Said Han Qingduo,"I am often beaten cruelly. There are a lot of scars on my body and please examine them now" (Deposition,717). Liang Ayou explained his obvious deformity as follows: "Once I left the store and didn't work for a day, and when I came back, the owner unexpectedly locked me up and cut off one of my ears" (Deposition 851). Bu Ahou witnessed the use of attack dogs upon a fellow coolie: "The white manager and the white overseer asked four dogs to bite his feet; his feet were bitten until they festered and he could no longer walk anymore" (Deposition 808). Wang Changtai described twelve of his peers who died "because they were flogged so hard that their flesh festered and had maggots." Another described how the overseer seized a knife and chopped off four of his fingers, and another described how boiling water was poured on him (Depositions 1035 and 1139). While powerful and startling as descriptions of domination and resistance, even more, intriguing is what these testimonies suggest in terms of the economic models of "transition" from slavery to a free society. The nature of these episodes underscores the imperatives of *retrenching* a particular racial order, especially when coolies threatened that order. Coolies continually reminded the slave-owning class of the pressure to transition from slavery to wage labor or free labor. This was not only an economic arrangement but a cultural arrangement as well. As Orlando Patterson and Demetrius Eudell have emphasized, slavery was a cultural institution, and, the ending of slavery and the project of "emancipation" was not only a "transition from slave labor to wage labor" but could be "described in terms that understand slavery as a system of social relations—indeed, as a cultural system." The most visible embodiment of transition in Cuba was the coolie. Thus, coolies had to be kept in check, subdued, and violently marked. Brought directly into a culture and labor system of race-based slavery, coolies were nonwhite figures—they were neither white, black, nor mulatto—who limned the possibility of becoming a free and "foreign" force. Coolies represented a new challenge to colonial authority, Catholicism, creole culture and language, and a race-based hierarchy of slavery. Commentary from a prominent planter at the time, Domingo Aldama, illustrated local sentiments regarding Chinese as potentially problematic in terms of maintaining a certain cultural and racial order. He argued that introducing *el chino* to Cuba would cause problems in terms of "race, color, religion, and sentiments" and further warned against their adding yet more potentiality to black revolt. Given their demonstrated rebelliousness, the Chinese coolies did raise the specter of insurrection, and furthermore, stoked the potentiality of alliances between the coolie and slave. In the midst of the thirty-year war for Cuban independence, increasing external and local pressures toward abolition, and technological changes in sugar processing, the Chinese were not only a disruptive force but also *represented* "transition," a historical moment that they described as a micro-struggle of violence.

THE PECULIAR FATALITY OF COLOR

The Chinese were a disruptive force at every turn, just as they were hypermobile from site to site. For example, Wang Dacheng, who "worked in many places," was transferred twelve times over a ten-year period (Deposition 241). At the macro and micro level, the mobile coolie continually retested the boundaries of a worksite and its social, cultural order, and confronted what one writer called "the peculiar fatality of color." Marked as inferior subjects, the Chinese described racist attitudes that foregrounded their subjugation and compared themselves to black slaves having white owners. Expressions in the testimonies stress that both Africans and Chinese were relegated to the lowest social strata. The depositions contained similar observations. One coolie noted that "when I first arrived here, there were six hundred workers, who were black and Chinese. Now there are only four hundred workers left. The black people work the same as Chinese, who are beaten and locked up as well" (Deposition 643). Another simply concluded, "On the sugar plantation, they treated black persons and me in the same way" (Deposition 244). Another coolie observation went as follows: "Black and Chinese are locked up in the same cell. At night, we have to wear shackles and in daytime we wear foot chains to work," and as yet another succinctly put it, "two cells, one for black persons and one for Chinese. And both blacks and Chinese in cells wear shackles" (Deposition 231). The Chinese perceived that they and Africans were placed in similar predicaments of inferiority. One statement by Chen Aer encapsulated a mapping of racial economy: "White men treat us Chinese as black slaves" (Deposition 216). The "white men" and the "black slaves" were the bases of comparison and polarity in the spectrum of racial power within which the Chinese were inserted. However, the coolies perceived themselves as more like black slaves than white men. On this point, James Williams' Steele, American consul in Matanzas, made an interesting observation regarding Chinese, Africans, and the hierarchy of color:

> When a negro who has killed an overseer one morning is led out and shot the next, when you may go out in the street and take the census of the chain-gang and find in one division of it sixty China-men, eighteen negroes, and no white men, and when you know at the same time half a dozen men who have testified absolutely to the intentional and premeditated killing of people before the fiscal, and know that the subjects of this uncontradicted testimony went free, and when this kind of thing passes under your observation for years, and nobody ever denies it, and everybody considers it a matter of course, it begins to seem as though there were a peculiar fatality in color and accompanying poverty.

Of course, the "peculiar" fatality in color and accompanying poverty was not unique to Cuba. But moreover, Steele's observation suggests the preponderance of Chinese in the lowest social strata. The coolies consistently described themselves as far from the owning class, which in their depictions consisted of white men, though one also mentioned a white female owner (Deposition 846). They located forms of ultimate authority as embodied by whiteness, commonly referring to "the white man," "the white manager," "the white owner," "the white official," and some cases, the "owner's wife." In no instance were Africans described by the coolies as owners. Rather, the testimonies indicate the particular feature of black overseers assigned to oversee coolies. The Chinese referred to "black people" as "slaves" or "black overseers." As one coolie explained, "Black people are slaves, who don't get paid" (Deposition 729). Furthermore, Chinese made note of the generational aspect of African slavery, pointing out that offspring of coolie-slave unions would also be slaves, as one Cai Xiang exclaimed, "I saw some Chinese have children with black women, and their children will still be owners' slaves" (Deposition 1054). Another Chinese laborer recollected a fellow coolie who married an African woman and pondered whether their child would be a slave (Deposition 1081). These brief mentions of coolie-slave unions, unfortunately too brief to draw further conclusions, revealed the ambivalence of coolie-slave relations as conflicted yet enmeshed. Some testifiers emphasized this ambivalence when they divulged covert collaborations between slave and coolie, despite overtly antipathetic acts against each other. The account by He Lin underscored the covert passing of information:

> One day, the overseer beat Li De to death in the sugarcane field and buried him there. Four blacks assisted him, and even officials did not know it. I did not see it, but the blacks told me. I have been in the sugar plantation for three years. Then nine of us made up our minds; we beat the overseer to death. We were sent to jail in Havana. We stayed there for two years and then were sent to a sugar plantation. We worked there for four months. Then the manager heard that we had beaten an overseer to death, so he sent us to jail in Sagua la Grande where we stayed for another two years. (Deposition 769)

In another example, Lai Axi recounted an African slave who sabotaged the Chinese, yet later shared information about the incident secretly: "One night when we were pressing sugarcane, a black slave put an iron in the sugar refinery machine and accused of us trying to break the machine.... Later, the black slave told me secretly that the manager told him to put that iron in the machine." (Deposition 216)

More often than not, however, the Chinese often regarded their social status as similar to that of slaves; but sometimes they felt less privileged than slaves, and therefore of lowest social status. Period tracts of scientific racism offer a counterpart to the testimonial depictions of "lowest" social status. One Cuban medical doctor, Benjamin de Cespédes, issued a study (1888) that ranked the races of Cuba according to sexual behavior and prostitution. This study ranked the sexual relations between Chinese at the very bottom (below prostitution between people of color (black), minors, and men). Apparently the doctor witnessed prostitution between Chinese men, which he deemed the ultimate act of sexual "abomination" and innate racial "inferiority." Relegated as a group to inferior status, coolies also described experiencing their oppression in keenly racialized terms. "Black and white here all treat Chinese badly," said Wu Axiang. "They always kick us, if they see us walking on streets" (Deposition 114). According to coolie accounts, racial hierarchy was reinforced in daily treatment but most keenly felt in terms of daily necessities and privileges that were granted or denied. Most of the coolies would not survive to see generations of descendants, let alone survive the next month or year. The "mundane" was described in terms of struggle. Getting sleep, getting food, and the preponderance of death emerged as overwhelming concerns. The Chinese repeatedly described being hungry, and in some cases, they described the starvation deaths of other coolies. "Some [Chinese] got sick and died because they did not have enough food to eat," declared Liu Guangcai of the Conception plantation (Deposition 727). "I am always hungry here," said He Asi, "and I have been beaten a lot. I am treated worse than animals" (Deposition 376). Aforementioned Huang Shirong spoke of a coolie who desperately sought something to eat: "He was so hungry that he ate sugarcane in the cane field, and when the overseer spotted him eating the sugarcane, he was beaten again immediately" (Deposition 670). Likewise, Lin Abang observed, "If people were caught for taking a little food during working time in the sugar plantation, they were forced to wear shackles" (Deposition 174). One Chen Long described how he snuck sugar water to sustain himself when working in the mill or refinery, "When I make the sugar,

I can drink some sugar water to allay my hunger; if I do not make the sugar, I have nothing to eat" (Deposition 977). There were 709 deponents who described being in a constant state of hunger. And 953 deponents described being surrounded by death.

The examples of Xu Shaolin and Pen Wendao are most interesting because of their positions as cooks on plantations—a position that would have allowed the *most* access to food. Pen exclaimed that the Chinese "who work on the sugarcane field can't have enough food and are hungry all the time" (Deposition 829). Xu Shaolin described not only his experience with food deprivation but also the role of food in *racial division*:

> There used to be forty-seven Chinese in the group, and now only fifteen are left. We are treated badly. They beat us and do not give us food... . Although I work as a cook, I cannot leave here even for a moment. If I leave only for a few seconds, they will beat me, just like the laborers who plant sugarcane on sugar plantations.... . Although I work as a cook, I still do not have enough food to eat. Even though some food gets left, it is given to the black people and not me.... People in Cuba already got used to enslaving the black slaves, and they treat Chinese worse than black slaves. This does not make any sense. (Deposition 855)

Starvation, much like death and burial rites, was experienced by coolies as inseparable from the politics of race-based privileges. In those terms, coolies compared themselves to slaves, and ultimately concluded that Chinese status was more in line with dogs. Death was featured in the testimonies as not only related to daily survival but also to cultural practices, as Chinese believed that the denial of burial rites led to eternal unrest. The concern with burials (and food), inextricable with the racial context in which deprivation emerged, was a node of racial strife. The coolie observations on this point are best exemplified by a coffin maker, Guo Amei, on the Flor de Cuba plantation, who declared the following: "Here dead Chinese won't be buried with coffins, but blacks are. I know, because my job is to make sugar cases and coffins" (Deposition 713). Another coolie made a typical observation: "I see the Chinese workers do not have a coffin upon death. They do not allow people to ask and do not allow friends to see the dead off. He (the owner) will cry if his puppy dies. We are truly not better than the dog" (Deposition

694). Chinese compared their treatment to that of dogs; but more often they felt they were treated worse than dogs, since dogs were at least mourned by their owners. The Chinese compared themselves unfavorably to slaves because they witnessed slaves being buried in coffins (though one might question whether some slaves, or perhaps less privileged slaves in the social hierarchy, might not have been discarded as the Chinese were). Due to the importance of ancestral worship and the importance of clan lineage in Chinese culture, the successful completion of funeral rites marked the critical moment of cultural and generational continuity. Elaborate burial rites and proper placements of burial sites in Chinese tradition involved practices that had been in place for thousands of years. The deep sentiment attached to the funeral emerged not only due to cultural practice, but also as a practical index of social importance and status. The lack of rites, and moreover, the manner in which this occurred, created deep-seated resentment among the Chinese for what they understood as their debasement as a group. Even baptism and conversion to Catholicism apparently did not ensure proper burial. The coolie testimonies often ended with the embittered refrain,' "Chinese are buried like dogs." Testimonies by Huang Afang, Li Achang, Wu Zhangsi, Pan Wendao, Lu Zelin, and Chen Amu, provide examples of common observations made in the overall testimonies:

> When the Chinese died, they were buried like dogs. (Deposition 413)

> If Chinese died here, their bodies would be discarded into a big hole, which is not far from the jail. And that hole is especially for dead Chinese. (Deposition 523)

> I witnessed Chinese who were buried like dogs without coffins. (Deposition 577)

> After Chinese workers die, their bodies won't be buried in graves. Instead, their bodies will be just discarded around. We are not allowed to help bury those dead Chinese. (Deposition 829)

> If Chinese died, their bodies would be discarded into a huge hole. The hole can contain four, five people and sometimes more than ten... And later the hole would be covered with mud. (Deposition 864)

> The food here is not enough, and we have to work every day from three in the morning to twelve at night with only two hours for

meals and rest... And we even have to work on Sundays... If Chinese died here, they will not be buried in coffins. (Deposition 905)

Coolie experiences were steeped in the racial divisiveness of slavery, with Chinese and Africans being pitted against each other. Wen Azhao observed that "black people were asked to drag and discard these dead bodies" (Deposition 50). Or as the aforementioned He Lin described, "four blacks" were recruited by the overseer to bury a coolie (killed by the overseer). The treatment of coolie deaths emerged in ways that emphasized the coolies' low status yet also revealed the white recruitment of enslaved Africans in the process of Chinese death and disposal, further reinforcing and antagonizing racial conflict.

STRUGGLE BEFORE SOLIDARITY: "ONE DAY WE COULDN'T TAKE IT ANYMORE"

Did the Chinese coolies and Black slaves transcend the racial divisions in their struggles for freedom? Is there evidence of a Walter Rodneyite vision of crossracial solidarity arising from shared oppression of labor? One simple answer would be "yes": There emerged solidarities, especially in consideration of covert and collaborative resistance in bondage, of Chinese and Black insurgencies that formed in the wars for independence, and of interracial social relations, of which some Afro-Chinese descendants are among famous figures in Cuban cultural and political history. Juan Jiménéz Pastrana's important work has underscored the significant roles of Chinese *mambises,* and a recollection of twentieth-century Cuban revolution by three Chinese Cuban generals acknowledges the significance of an earlier history of Chinese insurgence. The 1874 testimonies, however, revealed the underside of solidarity and the ugliness of racial conflict in the context of slavery. Unlike Rodney's history of the working people of Guyana, in which he recounted Asian-African conflicts as not characterized by fatalities, the micro-history before and during the solidarity between Asian and African people in Cuba was raw and brutal, of fatal intensity and human cost. The coolies' representations of interracial conflict disturb commemorative renderings of third-world solidarities, and they expose a more complicated and challenging history of struggle in the context of multiracial bondage.

In the twist of contract labor married to slavery, coolies resented slaves for having higher social status, and slaves resented coolies for being "not slaves." The following excerpts from Gao Alun and Yu Yecheng, respectively, provide examples of the grim consequences of racialized polarities under conditions of extreme exploitation.

When I arrived at the sugar plant, there were one hundred Chinese there. After eight years, there were only thirty left. Most of them died.... After fulfilling my contract, the owner detained me for four months in order to make up the time I used to be lazy while working. After the four months, he detained me for another three months. He declared he would give me twelve dollars a month. But he only gave me four dollars total in the first month and eight dollars the second month. So I refused to work. The second day, I and another nine workers went to a local officer, but he said we were runaways and sent us back. Back at the sugar plant, we were tied up in the middle of the night and beaten. The white overseer didn't stop lashing us until he was too tired to hit us anymore. I still have scars all over my body, which can be examined. Then we were locked up, and each of us was watched by a black person, who ordered us to cut grass. The grass was very tall, and we couldn't walk fast because of being chained; so the black people beat us. And then two of the Chinese thought it was hopeless, so they used knives to stab two blacks. One of the black people died, and the Chinese were beaten by the blacks and were wounded. The two wounded Chinese were then sent to the ward; we eight were still left cutting grass. One of the Chinese was beaten so hard and seriously wounded, and he died the next day. Then the owner asked the local officer to come and escorted seven of us to the yamen. The local officer asked me about my injuries and treated me. After two days, I was sent to a sugar plantation somewhere else. I have worked therefor twenty days with no payment and bad food. Then I refused to work because in my heart, I did not feel I should be punished. Later, I was sent to the official workshop in Matanzas ... (Deposition 536)

After six years, there came a new manager who was extremely brutal. So the blacks and some newly arrived Chinese worked together and killed the new manager. The owner spent some money and told the officer that the blacks didn't kill the manager, but instead said we ten Chinese, who were about to fulfill

our contracts, did it. So we were sent to a jail in Cardenas. One
of us refused to admit to the crime, so he was beaten to death.
(Deposition 612)

Such coolie testimonies described conflict as a life or death struggle steeped in racial politics. In Gao's account, a white overseer punished coolies for disobedience by putting the coolies under the control of black slaves, an arrangement which exploded into a fatal conflict. Yu, on the other hand, described the collective action of Chinese and Africans; yet, the integrated group became divided along racial lines. As a third example, the aforementioned Lai Axi told his story of a vengeful white manager who punished the Chinese for flaunting authority by pitting Chinese against Africans. Such tales exposed the intimacy of racialized antipathies with racialized complicities. In each recounting, the exercise of domination was channeled through the division between the Chinese and Africans. Plantation owners and managers undermined Chinese and African actions of solidarity, as in Yu's account, and disposed of troublesome Chinese as a method for destabilizing racial alliances and solving conflicts cheaply. Thus, the Chinese were disposed of for reasons of economic efficiency but in ways that were racially defined and divisive. "Economic efficiency" was revealed as interwoven with racial coercions and control therein.

In the schema of colored power, "black" was portrayed by coolies as the *in-between*, the middle, and the racial divide between yellow and white. The slave was rendered as intermediate between contract labor and ownership. This coolie perspective that "people here think Chinese are worse than black slaves" (Deposition 144), is an inversion of racial narratives of the Americas, which foreground the black-yellow-white and slave-migrant-ownership progression as constructed through the lens of immigration, modernization, and liberal politics. At the same time, the coolie depictions of slave-overseeing were still couched in terms that highlighted ultimate authority as consistently white. Most obvious in the testimonies was the coolie struggle against-white authority. You Asi perceived the corruption of power as stemming from the owner as evidenced by the following testimony:

On my way to sell fish, I was kidnapped by four people to a small
boat. They hid me in the bottom of the boat and took me to Heji
Pigpen in Macao. I was given a contract and one foreign dollar.

In the eleventh moon of the ninth year of Xianfeng (December 1859), I was shipped to Havana and I stayed there for two months. Afterward, I was sold to a sugar plantation in Colon with forty people. But after seven days, one man was beaten to death. His name is Chen Agou and he was from Fujian. He was beaten because he was sick and weak in the feet. I was hit a lot. One day I was beaten so hard that I even passed out for a while. But later I came back to life. We worked from three in the morning to eleven at night. After eight years, except me only nineteen people survived... Three people from Nanhai hanged themselves, and sixteen people died of illness ... The owner asked managers and overseers to beat me. He said, "even if one is beaten to death, I still have money to buy another ten people. " He was really mean and evil. (Deposition 240)

You Asi observed that the owner "asked managers and overseers to beat" him. Another coolie, Chen Qiguang, remarked upon subordination under a racial hierarchy of owner, manager, and overseer: "The owner treats us horribly; the black overseer and the white manager hit us a lot and torture us as much as they want" (Deposition 719). In another example, Lai Yasi indicates how the owner espoused brutality and promoted the perception among Africans that Chinese were disposable: "I remembered two years ago a Chinese was beaten to death and his body was discarded into the sea. Then some foreign soldiers saw his body and drag it to shore. After investigating, they knew this man used to work in this sugar refinery. Although the owner denied it, we all know he used to work with us. But the local officer didn't keep investigating the incident. The owner often said to the black overseer, "you can just beat Chinese laborers as hard as you want; if one is dead, I can always buy two more" (Deposition 812). Chinese-African relations, dialogically constituted within a specific history of extreme exploitation, necessarily included the "dialogic third" of the white owner, whether actually present in the testimony or not. In fact, the "owner" was not always physically present on the plantations, as one coolie noted, "Black overseers also often, beat people fiercely. Someone gets wounds all over his body. The owner comes here once or twice every year. In the sugar plantation, the manager takes charge in everything" (Deposition 977). The social order of the plantation, as depicted in the testimonies, emerges as some combination of the following: white owner, white manager, white overseer, black overseer, Chinese overseer (rare), black slaves,

Chinese foreman, and Chinese coolies. Beyond the owners, managers and overseers had great influence over whether a coolie would be disposed of and how. The coolies made references to both white and black overseers, describing the relationship between coolie and overseer as fraught with tension. Coolies went to lengths, however, to further elaborate on their black overseers, remarking upon especially brutal relations between them. As Yuan Aishan put it, "Black overseers are the most brutal ones" (Deposition 378). Due to the racial economy of slavery, Chinese would have viewed their being placed under Africans as a more humiliating subordination, being dominated by not only white owners and managers but also working under black slaves and overseers. In the social order of slavery and in the ideologies of white supremacy, black persons were considered in slave society as inferior. Thus, in the chain of power, the placement of coolies under black persons signified being reduced to "less than slaves." Given the hierarchies of race on the plantation, owners utilized this particular mode of racial subordination as a form of discipline and punishment, such as in Gao Alun's aforementioned testimony, which described being placed under a black overseer as further punishment for refusal to work: "The white overseer didn't stop lashing us until he was too tired to hit us anymore. I still have scars all over my body, which can be examined. Then we were locked up, and each of us was watched by a black person, who ordered us to cut grass." Through this racial lens, an even deeper resentment was manifested between two subjugated groups, with Chinese resenting the black overseers who had power over them. Couched in the context of a mixed labor economy, the Chinese were confounded by the social contradiction of their labor assignation as "contract labor" versus their social reality "as less than black slaves." The particular resentment against blackness is suggested in certain forms of expression, as some testifiers framed their remarks as rhetorical questions rather than as statements: "Why" should blacks be better than Chinese and "why" should "we" be less than "they"? Other remarks indicated perplexity, such as "it doesn't make any sense" or "have you ever heard of such a thing." Thus the increased rage of coolies against black overseers came about partly — or precisely — because they were elevated as a group to positions of superiority over Chinese. In turn, a black overseer likely harbored resentment of coolies as a nonslave, nonwhite group with a supposed wage. Simultaneously and crucially, black overseers would have had to *prove their (racial) fitness* for the job by forcing the Chinese into submission. Some illustrations of this are as follows:

A black overseer, named Phillip, hit us ever more severely. After more than three years, nineteen of us were beaten severely by Phillip and hated him so much. One day we couldn't take it anymore and killed him with hoes. (Deposition 432)

The black overseer was really brutal and hit us all the time. Once four of us couldn't take it anymore, so we killed the overseer. We knew if we didn't kill him, he would beat us to death in the end. Therefore, we did not feel bad about killing him. Four of us were sent to a jail in Colon for sixteen months; and then we were sentenced to jail in Havana for ten years. To me it is equally miserable to be in jail and in the sugar plantation. . . . I'd rather die than stay in Cuba. When I killed the overseer, I thought they would sentence me to death. (Deposition 574)

A black overseer hit us a lot. Four of us decided to kill him, and then we stabbed him with knives. Later we came before the local officer and gave ourselves up. The officer sent me back to the sugar plant and asked the owner about the situation. Later during the trial, I admitted the crime and was sent to a jail in Colon. (Deposition 550)

After more than one year, the black overseer was becoming more outrageous. We five people would rather give up our lives than be bullied by him any longer, so we killed him. (Deposition 312)

We are treated poorly on the plantation. Four black overseers hit us all the time. I have been beaten and worn shackles four times so far. Two of the times, I didn't make any mistakes, but the overseer reprimanded me. I tried to explain to him, but the overseer got upset and locked me up, which meant I wasn't allowed to talk back. I was also locked up once for leaving the plantation and another time for feeling sleepy at work I worked with another thirty Chinese, but only eight of them are left now. (Deposition 690)

In the black overseer/Chinese coolie arrangement, one subjugated person or group had to assert and maintain authority over another—although always under the auspices of the dialogic third, a white owner. Take the case of one black overseer, "Phillip," who was charged with overseeing nineteen (and perhaps more) Chinese, a situation ripe for explosion. And the last example depicted black oversight as not only a single overseer but as representative of a racial stratum of power over Chinese. The executions of coolies for killing black or white overseers further underscored the racial hierarchy. The respective testimonies by Li Qi and Lin Asheng provide examples of the conflict this hierarchy created:

> Once four people killed a black overseer, and were arrested and locked in cells in the sugar plantation. After six months, two of them were hanged; two were shot. Every worker witnessed that. I don't know whether officials have investigated that or not, but in fact, none of us workers have ever been asked to show up in court. (Deposition 345)

> After five months, six of us beat a white overseer to death with hoes. We were sent to the local officer; three of us were shot. I and two other guys were sentenced to jail for ten years. (Deposition 955)

The testimonies concerning Chinese subordination and the cleaved relationship between Chinese and Africans, particularly in the coolie-overseer arrangement, complicate a romance of interracial solidarity and its convergence with revolution. While not dismissing the resistant alliances that were eventually formed, both on the plantation and in the long-lasting insurgencies for independence, this reading of conflict complicates a progressive narrative of resistance and transradial solidarity, and highlights the conflicted nature of interracial relations formed under bondage. The coolies conveyed a consciousness of their collective lot as subordinated labor in a racialized context, with many of those incidences being subordination under African slaves or overseers.

A reading of interracial relations becomes further nuanced with a reading of *Chinese* intraracial conflict. The Chinese labor force was fraught with intragroup conflict when white owners elevated Chinese into positions of punitive power against their own. Chinese foremen emerge sporadically as prominent figures in the testimonies. They were occasionally mentioned in the position of the lead coolie, as foremen who functioned mainly as translators because of an ability to communicate, however marginally, with the overseer or manager. In a handful of instances, Chinese foremen were characterized as reasonable. Yet in other cases, such foremen could assert themselves in brutal ways. In those cases, it might be assumed that the Chinese bore the abuse by Chinese foremen because, of cultural commonalities or racial solidarity; but the testimonies revealed that, in fact, the Chinese resented —and sometimes murdered—their Chinese foremen. Thus, the exploitation system functioned not only by turning Chinese against Africans but also by turning Chinese against Chinese. Conflict took place not only against black overseers but also against Chinese foremen. For example, note the following testimonies:

> When I first got here, I was beaten a lot. Later I was assigned to be a foreman. I didn't want to hit people, but the manager asked me to do so all the time. If I didn't hit Chinese workers, I would be locked up and beaten by the manager. As a result, I hit lots of workers. Once while I was asleep, a worker, who was beaten by me before, chopped my face. I was almost dead. (Deposition 555)

> Last year we had a Chinese overseer in the sugar plant. He treated people badly and was killed later. (Deposition 977)

> The Chinese overseer often beat us....Six of us beat him to death. (Deposition 897)

> Once a Chinese overseer from Fujian asked a white to lock me up, I stabbed him with a knife. I stabbed him twice in the neck and twice in the chest, but he didn't die. (Deposition 864)

The perception of racial and cultural betrayal is suggested in the Chinese-Chinese relationship, as the Chinese overseers survived by exploiting their "own," and in those instances, descriptions were of Chinese turning against Chinese. The testimony by the Chinese foreman, Zhang Yi, revealed his assignment to be racialized; he was promoted exactly because he was Chinese and because he was deemed capable of meting out punishment *against* his "own." Like black overseers, the Chinese foremen were compelled to preserve their positions and to survive as long as possible in their circumstances. The racialized politics of "selling out" in a system wherein survival turned upon exploiting one's "own," is further contextualized by the repeated, embittered accounts of coolies who testified that they felt betrayed by fellow Chinese who initially sold them to white foreigners in coolie traffic. On three fronts of Chinese-Chinese, Chinese-African, Chinese-White relations, the testimonies suggest complicated motivations for dominance and resistance, and ethnic schisms within the coolie and slave labor force. Notwithstanding the basic battle to survive, their depictions of resistance were invariably contextualized by the racialized mind-sets and practices that created their bondage to begin with.

A consideration of racialized mind-sets raises the possibility of cultural preconceptions of blackness and slavery embedded in Chinese cultural views *before* their arrival in Cuba. In the entire body of testimonies, none indicated or referred to relations or encounters with black persons before arrival in Cuba. In China, neither a system of generational black slavery nor a mass migration of African people had existed. Still, it is entirely possible that the coolies were aware of images of black slaves and racial discourses regarding slavery before exposure to black slaves in Cuba. Not enough has been written or explored on the subject of black slavery as it was viewed in China, though recent studies indicate that early African-Chinese encounters included two different patterns: One was of reciprocity and mutual respect, and another was of slavery. In the extraordinary history of Chinese navigation overseas, Louise Levathes documents epic and extensive court-sponsored expeditions of the Muslim Admiral Cheng Ho, or Zheng He, who undertook his voyages from 1405-1433. At one point, his enormous fleet arrived on the coast of Africa and engaged in a respectful exchange, and even fulfilled one mission of safely returning African emissaries to their land. The African emissaries had earlier ventured to China and presented gifts to the Chinese, including African animals such as the giraffe. Early Africans thus arrived in China as diplomats on one hand, and on the other, as slaves. There is evidence that small numbers of African slaves were introduced to China via Arab traders. Janet Abu-Lughod discusses the cosmopolitanism of southern China at this time and the presence of people from various parts of the "world system," not the least of which were the Arab traders who formed a "sizable Muslim merchant community." The traders brought slaves with them and installed them as domestic servants. Subsequently, there are indications that dark-skinned slaves were posted as guards to homes of the wealthy Chinese. In Chinese literature, dark-skinned slaves emerged called K'un-lun-nu, with Kunlun becoming a popular figure in *the Legend of K'un-lun-nu*. A tracing of the Kunlun history and its representation from the Tang era has been undertaken by Victor Cunrui Xiong, who noted that "Kunlun" in fact refers to the Malay peninsula, where dark-skinned people lived. Still, he concluded that Kunlun slaves of the Tang, in fact, referred to people from Africa.

This literary figure and its relation to Chinese attitudes reemerges in recent examination of African-Asian histories. In her long view of the history of Africa and global interaction, Maghan Keita recently examined African-Asian and African-Chinese interactions and the historiographic literature that exists on the subject(s). She noted that the presence of Africans in China may be registered in three ways: "African artifice and goods, the African and Africa imagined, and Africans as the people they were and are." In her analysis, she is in agreement with Philip Snow, who produced a monograph on early African-Chinese encounters, and Samuel Wilson, who also examines the subject, albeit less extensively. Keita referred to the black Kunlun figures that emerge in the literature of the Tang era (618-906 C.E.) and, according to Snow, are portrayed in a specific context: "[They] speak Chinese, behave like Chinese and are treated by their Chinese owners with every sign of respect. The Kunlun are no common servants. They are unfailingly heroic and resourceful." Yet Keita, in agreement with Snow, asserted that a shift in attitude occurred, brought on by the increased presence of Arab merchants and slaves, and that "the values that had informed the Kunlun—the African, the black, the slave—shifted as well. They came to be characterized in the same manner that masters regarded most subject populations: servile and, many times, subhuman." Still, it is unclear when, how, or to what extent this shift occurred. The Zheng He voyages were abruptly stopped during the Ming Dynasty (1368-1644 C.E.), a ban was implemented regarding maritime trade, and relations with foreign powers were halted. This was a reversal of such magnitude that scholars today are still pondering the historical reverberations. Any traffic in African slaves would have ceased. Given Keita's useful survey of historiographic sources on the subject, the relatively slim body of scholarship on African slaves in China and the extent of their presence, the lack of evidence specifically locating Chinese attitudes in this regard, it is difficult to define the pre-Cuba "roots" of Chinese coolie attitudes toward Africans. However, it is possible to infer Chinese attitudes toward dark skin and thus, blackness. Dark-skinned slaves in China included dark-skinned Chinese, Malays, and Indonesians. Historical studies indicate that all groups were utilized as servants and slaves. Serfdom in Chinese society was hereditary and perpetual and was in place until its dissipation in the eighteenth century. In this respect, "blackness" or dark-skinned labor was associated with bondage and peasant labor, and notions of perpetual servitude were naturalized in Chinese culture. Chinese systems of bondage and serfdom carry their own historical particularities and characteristics that cannot be homogenized with the Atlantic slave system, though it could be said that the Chinese cultural history of bondage and that of dark-skinned peoples, combined with vestiges of the "Kunlun" slave, may have informed Chinese with notions of "blackness." Coupled with associations of dark skin with manual labor, there was the Chinese tradition of the gentry and aristocracy being marked by non-manual labor, less exposure to sun, and fair skin. Furthermore, nineteenth-century European and American imperialism in China, and the coolie traffic, was conducted under long-running discourses of race and colonialism. The Chinese populace was outraged by the traffic and public riots ensued (see Chapter 1, "Historical Context").

Public denunciations of the coolie traffic involved comparison to African slavery, including denunciations issued not only by Chinese but also by Westerners in China. One journal of the period, edited by W.A.P. Martin, declared the following: "Even though Chinese are hired as contract laborers; in fact, they are treated as slaves with no difference from the blacks of Africa." Therefore, testimonies from Chinese in Cuba offer no direct evidence of their preconceptions regarding blackness; yet the nature of "attitudes" is that they are naturalized into acts, texts, and expressions. The cultural and material histories of slavery and serfdom, the history of both Chinese and non-Chinese slaves, and the impact of European attitudes could have served as pretext to racial perceptions in the 1874 testimonies.

THE COST OF DOMINATION: "THE CROWD FOUGHT BACK"

A prominent feature of labor histories concerns the cost of production. Most striking in this history is not only the cost of production, but also the high cost of domination. Here, "cost" is put forward in terms of life and limb. Underlying this rather apparent interpretation of cost is an examination of domination, revealed as uneven, penetrable, and subject to daily resistance by coolies. There are 953 oral testimonies (or 81 percent of the total oral testimonies) that described the omnipresence of death contextualized by a numbing litany of punishment, retaliatory killings, and suicides. The coolies were cut off from their families in China, with little hope of returning. They were without enough food and lived with the dread of an unsettled death or no funeral, and most of all, there was a heightened arbitrariness and expendability to their lives. As a result, the coolie relationship to management was marked by volatility. The myth of return that arguably sustained immigrant "sojourners" abroad haunted the coolies in Cuba. But these coolies expressed the dawning realization that, in fact, they were permanently trapped and would never see home again. Thus, many felt they had nothing to lose: "How dare a Chinese worker dream of going back to China?" (Petition 2) This predicament of unending bondage combined with a high mortality rate led to the preponderance of explosive resistance and chaos. Luo Aji described the life-or-death struggle in the sugar field: "In the fifth year, I was sent to cut grass with Huang Axing and Chen Aye. Because we didn't cut one branch, the overseer hit Axing with a whip. Then Axing turned around, the manager tried to cut his neck with a machete. So Axing immediately took that machete and chopped off the manager's hand. Then Aye injured the manager's head with his sickle. I did not do anything, but I was also sent to jail with them" (Deposition 431). The implements of slave driving and hard labor made appearances throughout the testimonies in the form of whips, chains, clubs, axes, hoes, knives, machetes, and guns. This "weaponry of work" was not only wielded by managers and owners but by the coolies themselves. For example, the weaponry of work appears in Huang Azhang's account: "The overseer wanted me to bribe him, so he made me wear shackles and hit me severely with cowhide all the time, and sometimes he hit me with a club. Lin Ayou and I could not bear this humiliation and bullying, and besides, our lives were in danger, so we killed the overseer. I used an ax, and Ayou used a machete" (Deposition 447). He Afa also describes the omnipresence of slavery's implements and their violent use:

> There are two meals of corn everyday. We work from two in the morning to one at night. The time for us to eat and sleep totally is no more than five hours. If we got up slowly or walked too slow or too fast, we would get whipped. The whip is made of a big slice of leather, so using it to beat us makes our skin crack and blood flows out. There is a cell in the sugar plantation which is usually full of people with neck locks, foot locks, and shackles. I had an internal injury, so I was the one who most often got beaten. In the sugar plantation, I saw by my own eyes that Zhang Awang, who came from Guangdong, could not stand the torture anymore

and hanged himself. There were two people from Qiongzhou City in Guangdong who committed suicide by swallowing opium. I was beaten countless times. Once a manager who beat us relentlessly for days, took out a machete and wanted to slash us. Then the crowd fought back and slashed him to death instead. (Deposition 883)

Testimonies such as these depicted combative struggles in daily interaction and revealed the precariousness of controlling bonded labor. The candid and graphic nature of their words is surprising as is the pattern of remorseless and open admission, as it would seem to foreclose any possibility of liberation or redress before an official body. In these cases, these testimonies could be read as mass indictments of the system, rather than as pleas for help. The cheapness of contract labor and the easy replacement of that labor were manifested in disposability and mobility becoming methods of control over labor. Owners discarded coolies by weeding out resistant or unproductive coolies though punishment or death. Or, in some cases, coolies were simply passed off to the official workshops or other plantations and were cheaply replaced. The basic premise of control yoked "disposability" with "domination." One of the most powerful tools of control was the freedom paper system, which was manipulated by management and used to hold the coolie hostage to the potentiality of freedom. Yet, the high turnover and attrition of coolies and the seemingly endless deferment of freedom also produced enraged workers. Coolie Guan Ashou detailed their frustration, echoing testimonies of other coolies: "The overseer beat us so severely that some of us were beaten to death. Even though we fulfilled our contract, we couldn't get our freedom papers. So we'd rather kill the overseer now than be beaten to death by him later. It is much better to be put in jail than in the sugar plantation." (Deposition 504) Coolies fought back in situations marked by temporal compression, with testifiers stating that their time was limited and freedom was not forthcoming.

The life-or-death struggle and its component of fear pervade the testimonies: There was not only fear expressed by coolies but the fear that pervaded management. Managers and overseers, whose jobs were to maintain order and increase labor production, literally risked their lives in their struggle to extract labor from "voluntary labor." As one Zhong Jinxiu declared regarding his manager, "I

tried every means to kill him" (Deposition 679). The daily dealings
of management, as portrayed in the testimonies, reveal the strain of
attempting to control a mass group that was resistant. As one Zhu
Arui emphasized, "The manager used to treat us ferociously. Later
a few overseers were slashed to death, so now he treats us a little
bit better" (Deposition 877). While coolie labor has been understood
as a category of efficient labor used in a transition from slavery
to free labor, here the coolies reveal the resistance of such a force.
Authority was depicted as teetering on a line that was constantly
tested. In some cases, coolies described management as panicked.
Wang Mujiu recalled, in the following testimony, his overseer firing
madly into the crowd:

> We worked from four in the morning to one at night with two
> hours break for us to eat in between, but the food was always not
> enough. Because the work was so intense and toilsome, I reported
> the situation to the officer. The officer asked the owner to give me
> a smaller workload and enough food to eat. The owner promised
> to do so in front of the officer. But soon as we got back, he locked
> my feet with shackles for seven months. When I was in shackles,
> I had to continue working. Sometimes I was too tired to work,
> and he would lash me and lock me up again. I was asked to cut
> sugarcane. The overseer furiously beat me all the time. Once I
> could not stand it any more, so I stabbed his arm using a knife.
> Then the white man started to fire his gun madly, and a few people
> in the same group with me got injured. Afterwards, they sent six
> of us to prison in Havana. (Deposition 625)

One He Lin described how "nine of us made up our minds" to
kill an overseer and how a subsequent owner hastily passed them
off upon hearing of their crime.

In repeated examples, the Chinese decided with collective
forethought to rid themselves of particular managers or overseers.
Overt resistance included confrontation and sometimes fatalities,
but there were also organized assassinations of the overseer, the
manager, or both. Assassinations would involve the collective act of
a select few, or would involve twenty Chinese, or in some instances
a "crowd" of Chinese. Wu Zhangsi detailed the "deal" made among
his group: "Sixteen people hated the overseer a lot and we made a
deal; if the overseer hit us again, even if he just hit one of us, together
we would beat him to death. Then we killed him and all of us were

arrested and sent to jail in Colon. Three of us died in the cell, two were transferred to jail in Havana, and eleven were bought back by the owner.... Of those involved, four of us did not actually kill that overseer. But we were still locked up and beaten severely by the owner" (Deposition 577). And another coolie, Wu Jin, described the action taken by a group of ten: "Working for one and half years, I was hit a lot by a black overseer. One day ten of us decided to kill him with the machetes we used to cut sugarcane. Then we turned ourselves in and were sentenced to jail for ten years" (Deposition 920). Other examples of "assassination" testimonies include those such as Yuan Aying's, which described the mass act of twenty coolies who kill their manager: "His rule was we had to work day and night, and we had nothing to eat. I worked there for thirty-five days and could not stand it anymore. Totally twenty of us did it—we slashed the white manager to death" (Deposition 868). Mo Agang's following testimony recounted that five coolies "worked together and fought back" by bludgeoning the overseer to death:

> The owner and the manager were wicked people. The food was only potatoes and corns and which is not enough food to eat. When we need to press the sugarcane, we work from one in the morning to ten at night. If we do not need to press the sugarcane, we work from three in the morning to eight at night. In the sugar planta- tion, they use leather whips to beat us. After working in the sugar plantation for a year, I saw three people hanged themselves and five people swallowed opium to kill themselves. There were five people in my group—because five of us often got beaten by the overseers—we have wounds all over the body. One day when we were working in the sugarcane field, he started to beat us again. We worked together and fought back. We used hoes to beat him to death. The owner sent five of us to the jail in Cardenas, where we stayed for eight months. And then I was transferred to Havana and sentenced to jail for ten years. Now I have been in the jail for four and a half years. In the jail, I wear four different shackles. The officer in the jail assigned me to roll the tobaccos. I have to roll three packs of tobaccos, about sixteen thousand tobaccos. I have to work day and night in order to finish my 'workload, as I will get beaten if all the work is not finished. The food is only conjee, and I do not have enough food to eat. Working here, I suffer less wear and tear of the weather compared to working on the sugar plantation, but the humiliation is the same as on the plantation. (Deposition 868)

As punishment, Mo ended up in prison. Still, Mo's captors remained fearful of him, as he was required to wear four shackles while he did daily forced labor. The fear of "Mo" exemplified the fear of the coolie, who embodied the first social and economic experiment to mitigate the rising costs of slave trade, yet was clearly resistant to labor "efficiency." One could not point to a more extreme pattern of counterproductivity than workers retaliating against their managers and overseers, whether they were black, white, or Chinese. In some cases, there were workers on one plantation killing "several" overseers in sequence. Selections from other testimonies provide further details of violence and its contexts:

> The first manager was nice, but after three years, there came a new manager, who is white. He beat us all the time. Not being able to bear this anymore, a group of new workers killed that manager. (Deposition 868)

> The manager treated me badly, and I couldn't take it anymore, so I talked to another worker about how I had to kill the manager. One day the manager came and told me to work night shift; I killed him with a kitchen knife. (Deposition 264)

> Ten of us couldn't take his bullying and torture anymore, so we tried to kill him. But he did not die, only was injured. (Deposition 442)

> Five of us were beaten severely for two days in a row, and we were injured seriously on heads and faces. And our eyes were injured too. You can still examine our scars. We were so angry that we fought back and killed him. (Deposition 734)

> Nineteen of us suffered a lot. One day the manager again hit people for no reason; we tried to calm him down, but he tried to kill us with a long knife. So we fought back and killed him with hoes. (Deposition 873)

Yet the Chinese were not always executed for these murderous acts—obviously, there were those such as "Mo" who survived to recount their term of imprisonment. As one prisoner noted, "After three years, I and another two who also hated the overseer stabbed him to death with knives, because he bullied us a lot. But a white man witnessed this, and we were arrested and sent to the local officer in Matanzas" (Deposition 909). Given that testimonies underscored the need for control by management, it would seem illogical that violent resisters be allowed to survive. Resisters' lives were spared in part because coolies represented the possibility of profit *beyond* the onsite production. The testimonies regarding deadly resistance point to the overarching lure of profit. As flexible labor, Chinese were resold in a cycle of further profiteering. This recycling amounted to immediate profit for sellers of coolies and further production in the tobacco and sugar industries. Like Li and Mo's accounts (above), coolies were sent to jails as captive labor with high yields. Li and Mo described their quotas as fifteen to sixteen thousand cigars (or cigarettes) per week. Li Hui, earlier mentioned as one who "accidentally" killed his overseer, also provided an important glimpse of coolie prison labor in the following account:

> When we needed to press sugarcane, we worked from three in the morning to twelve at night with half hour of rest for meals. Other times, we worked until ten at night. After working for six years, one day at eleven at night I felt sleepy while cooking the sugar, so the sugar got burned. The overseer hit me with a handspike. I took that handspike and pushed him; however, the overseer fell into the sugar-cooking pot. I didn't mean to kill him. But the manager sent me to the local officer, who sentenced me to prison for ten years... Later I was sent to another prison in Habana, where I stayed for seven years so far. In the prison, being a tobacco-roller, I have to turn in fifteen thousand tobaccos a week to the officers. And I can earn one dollar by rolling extra five thousand tobaccos. The food in jail is not enough; however, I would rather stay in prison than in the sugar plant. In the sugar plant, I had to work all day and night long; the workload was heavy, and I was hit severely. (Deposition 67)

Han Yanpei, described a similar quota: "In the jail, those who are in the tobacco workshop need to make twenty-six hundred cigarettes per person per day" (Deposition 219). Over six working days, this quota would have amounted to fifteen to sixteen thousand per week. Over seven working days, this would have amounted to eighteen thousand per week. Thus, "jail" was a site of production in a chain of sites that contract labor circulated through. "Jails" and "workshops," with this kind of captive yield, and the liquidity of coolie labor and resale, need to be considered in analyses of contract labor and its uses.

The extraordinary details of the testimonies—from those young and old, of different work sites and locations, and from those who attested to their own experience but also bore witness to the circumstances of others—emphasize the experience of so-called transition from slavery as volatile, unstable, precarious, and brutal. Why was the experience of "contract labor" of such brutal nature? One most obvious reason was profit and the systemic pursuit of it—a motive most clearly exposed in the petitions. Perhaps Wang Guijie's account is the most chilling, as it matter-of-factly depicted profiteering from the coolie traffic: "I got sick when the ship arrived at Havana. I lived in a ward and then was sold to a clinic in Matanzas. The owner of the clinic always bought sick people and cured them, so he could sell them after they got better" (Deposition 481). Wang's offhand description of Chinese being healed so they could be sold describes not only the capture of body for the purposes of forced labor but also the colonized body as future investment and commodity. Suggested in this and the remaining testimonies were two apparently contradictory impulses that buttressed this first massive influx of Asians to this hemisphere, one being the imperatives of profit and the utility of the "raced" body, and the other being the fear of the racial differences that such profit thrived upon. The threat of a new racial body and the destabilization of a preexisting racial and social order resulted in the need to redraw lines of domination- and control. One Chinese Cuban of contemporary times recounts that, in the past, the Africans were called *niche* (equivalent to "nigger"), and the Chinese were called *narra*. In the testimonies, fear of the Chinese was accompanied by the impulse to divide Chinese from Africans—and in some cases, Chinese from Chinese. As a new social body, the coolies emerged and complicated an already conflicted social theater of *de color* when color was primarily defined along lines of blackness and whiteness.

The heightened intensity of conflict, and the compressed nature of shortened lives under disposable exploitation, raise singular questions about forms of resistance and "community" under conditions of expendability. The testimonies suggest the mass acts of testimony as constituting a political community in itself. The tropes of witnessing and naming, the desire to speak for the dead, the prevalence of group perspective ("we," "us," and "our"), and the narrating of collective actions, signal the forming of collective identity in the act of giving testimony. The petitions (particularly the mass petitions) emerge as extraordinary forms of collective resistance through writing, and the depositions reveal the highly charged acts of testifying against great odds, with some testifiers indicating that they spoke not only for themselves but for others who had no means of communicating. Repeated tropes and themes, frequently appearing phrases and refrains, common locations and contexts, locally formed coolie language to describe and make sense of a bewildering system—all contributed to the making of an overarching coolie narrative(s), rather than simply a massive assortment of unrelated statements.

Other questions remain as to how testimony-giving directly affected the coolies after 1874. How did this historical moment of testifying affect their subsequent perceptions and actions? How did it affect their views of themselves and their futures? What effect did this event have upon the coolies' subsequent relationship with cultural identity and nationalist sentiments? Was there any correlation to the numbers of coolies who continued to join the war for Cuban liberation? How did the testimony affect the views of local Cubans toward coolies, if it did? Is there evidence that links this episode of testimony to subsequent liberatory politics, both in Cuba and elsewhere? These questions remain unanswered here. For example, exact numbers and the tracking of coolies who joined the independence wars were, not recorded, though there is evidence of their mass participation. The testimonies and the commission report eventually led to the official end of coolie trafficking to Cuba but did not liberate coolies already in bondage. The last coolies in bondage would not be freed until the African slavery system underwent a slow demise in Cuba via a transitional patronage system in the 1880s.

Furthermore, there is no extant autobiographical or self-representational writing that has been left to us from the coolies, outside of their self-representations in 1874. As such, I do not suggest these testimonies as the über-source that represents experiences of all Chinese in Cuba at the time, especially as the history of Chinese in Cuba extends from the mid-nineteenth century until today and has various social genealogies. However, there is a fascinating piece of self-representation by a second generation Afro-Chinese author. He provided a narrative of the coolies as freedom fighters who emerged from slavery, but he ultimately portrayed them as marginalized figures of blackness amid the subsequent rise of a merchant class.

A new generation of Chinese Californians had arrived, and this would change the landscape. In this rare history of community, this author made the first attempt to comprehensively historicize the Chinese in the wake of bondage and did so in a transnational context. Dedicating his work to the coolies, yet paying homage to merchants and new immigrant success, he ambivalently cast coolie history under the shadow of new capital while also undermining dominant paradigms of modern immigrant success along the lines of race and labor. Ultimately, his work bridges the yawning narrative gap that opens after 1874 and offers a unique and vexing perspective to the question of what happened to the "coolie." We move now to the second generation and examine the vagaries of politics in the narration of "coolie-to-immigrant."

CHAPTER EIGHT

Indian Women and Labour Migration

VERENE SHEPHERD

Background to Maharani's Passage from India

Maharani boarded the *Allanshaw* destined for colonial Guyana in July 1885. By that year, the colonial Caribbean already had a long history of importing contract labourers from India for commodity production on its many, primarily sugar, plantations. The general history of this post-slavery Indian labour reallocation to the colonial Caribbean is well documented by K.O. Laurence, Hugh Tinker, Walton Look Lai, Basdeo Mangru and many others, and will, for the most part, only be rehearsed from secondary sources in this essentially background chapter which aims at locating and contextualizing the voyage of the *Allanshaw* and the experiences of those on board.

Colonial Guyana pioneered the Indian indentureship system in the Caribbean, receiving some 238,909 Indians between 1838 and 1917. This figure represented 55.6 per cent of the total of 429,623 Indian immigrants who arrived in the colonial Caribbean, and only Trinidad, which imported 33.5 per cent of the total number, came close to Guyana's figure. (See Tables 8.1 and 8.2.) As shown in Table 8.1, Indians outnumbered all other ethnic groups of indentured workers in post-slavery Guyana. According to Look Lai, there were three distinct phases of Indian labour migration to colonial Guyana. The first immigrants arrived in May 1838 on the ships *Whitby* and *Hesperus*.

These 396 arrivals were known as the "Gladstone Coolies", having been imported on the initiative of the sugar planter John Gladstone,

Table 8.1: Immigrants Introduced into Colonial Guyana, Mainly under Indenture, 1834-1917.

Source	Period of Immigration	Numbers
India	1838-1917	238,909
Madeira	1835-1881	32,216
Africa	1834-1867	14,060
China	1852-1884	13,533
Europe	1834-1845	381
Other	1835-1865	1,868
Total		300,967

Source: Clem Seecharan, *Tiger in the Stars: The Anatomy of Indian Achievement in British Guiana, 1919-29* (London: Macmillan, 1997), 3, taken from G.W. Roberts and J. Byrne, "Summary Statistics on Indenture and Associated Migration Affecting the West Indies", *Population Studies* 20, pt. 1 (1966): 127.

Table 8.2: Number of Asian Indians Imported to the Caribbean, 1838-1917.

Territory	Years	Numbers Imported
British Guiana	1838-1917	238,909
Trinidad	1845-1917	143,939
Suriname	1873-1918	34,024
Guadeloupe	1854-1887	42,595
Jamaica	1845-1916	38,681
Martinique	1848-1884	25,509
St. Lucia	1858-1895	4,354
Grenada	1856-1885	3,200
St. Vincent	1860-1880	2,472
St. Kitts	1860-1861	337
French Guiana	1853-1885(?)	19,296

Sources: K.O. Laurence, *Immigration into the West Indies in the Nineteenth century* (Barbados: Caribbean Universities Press, 1971); Gisela Eisner, *Jamaica 1830-1930: A Study in Economic Growth* (Manchester: Manchester University Press, 1961), 144; Seecharan, *Tiger in the Stars*, 4; Verene Shepherd, *Transients to Settlers: The Experience of Indians in Jamaica, 1845~1945* (Leeds and Warwick: Peepal Tree Press and University of Warwick, 1994).

who owned the plantations Vreed-en-hoop and Vreedestein. Gladstone himself used 101 of the 396 and the remainder were distributed to other plantations. This first phase was short lived and there were several stops and starts between 1838 and a temporary halt in 1848. Indian immigrarion did not resume fully until the 1850s, by which time some of its worst abuses had been corrected. In the second phase, 1851-70, Indians were in the majority but not the exclusive immigrants who arrived in the colony. The third phase, 1870-1917, however, was marked by the exclusive importation of Indian contract labourers. In this latter period, immigration to colonial Guyana, as indeed to Jamaica and Trinidad, was regarded as a new form of settler colonization pushed by British imperialist interests. While repatriation continued to be an integral part of the indenture contract, incentives were held out to entice the Indians to make the transition from transients to settlers.

The massive importation of Indians to the colonial Caribbean changed the ethnic composition of the plantation labour force drastically, especially in colonial Guyana and Trinidad. Whereas in 1851 the African-Guyanese dominated the plantation labour force, with Indian labourers accounting for just 16 per cent, by 1891 colonial Guyana's sugar plantation economy had come to rely almost exclusively on indentured immigrant labour, with Indians comprising 80.4 per cent of its 90,000 full-time plantation labour force. Up to the end of the Indian indentureship system, however, the African-Caribbean people in colonial Guyana, as in Jamaica and the Windward Islands, whether as part-time or full-time labourers, continued to be essential to certain tasks on the plantations. The abolition of that system of neo-slavery euphemistically called Apprenticeship in 1838, had given the African-Caribbeans a new mobility which was exploited fully. The trend was towards peasant formation and non-plantation occupations; few remained as resident estate labourers if they had a choice. Despite coercive planter tactics and other obstacles, a vibrant peasantry evolved in competition with the estates for labour.

As was the case with the shipping of African captives, the relocation of Indians to colonial Guyana and the wider Caribbean was characterized by a sexual disparity, especially during the early years of the scheme. There were only fourteen women among the 1838 Gladstone arrivals and this situation improved only gradually.

By 1856, 38.2 per cent of the 5,004 imported to colonial Guyana were females, well above the average of 16 per cent shipped from Calcutta to all colonies in that year. The records of the Colonial Land and Emigration Commission show that in 1857, female numbers reached a high of 69.6 per cent, decreasing thereafter to between 27.2 per cent and 47-7 per cent from 1858 to 1866.

The sexual disparity was not confined to colonial Guyana. Only 11 per cent of the 261 immigrants imported to Jamaica on the first Indian emigrant ship, the *Blundell*, comprised women. If girls are added, the female percentage increases to 15 per cent. In 1863 the *Alnwick Castle* to Trinidad carried only 14.6 per cent females out of Its 460 emigrants and the *Golden City*, of the same year, 13.4 per cent. The excess of males over females among Indian emigrants was also noted in Suriname where slavery was abolished in 1863 and immigration began in 1873. The 1872 treaty between the Dutch Parliament and Britain respecting emigration from India stipulated that a 50:100 female to male ratio was to be maintained. This was later lowered to 40:100 as it was for the British-colonized territories as recruiters could not meet the higher quota. This unfavourable female to male ratio, frequently below quota, led the British government to seek a workable solution by imposing standard ratios. These ratios fluctuated between the 1850s and the 1860s, being at various times 25 females for every 100 males and 50:100. The female to male ratio finally serried at 40:100 for most importing territories but despatching the requisite proportion of females for the colonies was a perennial problem for all concerned, as illustrated in the tables below.

Failure to fulfil the ratio requirements often delayed the departure of the ships out of the ports of Calcutta and Madras, but the social implications of a severe shortage of women, such as the high rates of uxoricide in the Caribbean, caused all concerned in the labour migration process to make the effort to come as close to the ratio as possible. As a concession against unforeseen recruiting difficulties, agents were allowed to send off a ship without the stipulated quota provided the deficiency was made up by the close of the following emigration season.

Recruiting agents continued after the 1860s to experience difficulties in attracting women to colonial Guyana and the other importing countries in the Caribbean. Not only were many women reluctant to leave friends and family in India but the system of child betrothal left few unattached women for emigration. Additionally, Indian men were reluctant to subject their wives, daughters and other female relatives to a long and potentially perilous sea voyage. Not all Caribbean planters were supportive of the emigration of "lower-caste" women, categorizing them erroneously as prostitutes.

Table 8.3: Percentage of Females Imported to Colonial Guyana, 1856-1866.

Year	Men	Women	Boys	Girls	Infants[b]	Total	% of Females
1856	—	—	—	—	—	5,004	38.2
1857	—	—	—	—	—	3,487	69.6
1858	—	—	—	—	—	7,566	47.7
1859	—	—	—	—	—	9,186	42.6
1860[a]	2,410	1,097	181	151	127	3,966	32.5
1861	2,180	817	340 ch.	_	97	3,434	27.3[c]
1862	2,352	601	94	77	92	3,216-	21.7
1863	1,509	374	48	31	52	2,014	20.6
1864	1,745	427	101	64	89	2,426	21.0
1865	3,071	845	297	161	282	4,656	23.0
1866	1,265	442	89	65	59	1,920	27.2

Source: CO 386, Letter Books of the Colonial land and Emigration commission.
[a] incomplete returns
[b] not differentiated by sex, therefore excluded from the calculations
[c] % of women, not females. Children not differentiated by sex
- = not stated

Some, influenced by the post-slavery Victorian ideologies which were being imposed on the Caribbean and ignoring the long experience of productive female labour during the period of enslavement, tried to hide this prejudicial sentiment by arguing that women did not make as good agricultural labourers as men, or that women should function in the private sphere. The restraints on the emigration of women who had large numbers of children with them also complicated the emigration process with respect to filling the female to male ratio. The view was that ships which took on board large numbers of children were likely to experience an outbreak of epidemic diseases such as measles and that the death rate was likely to be high in such cases. For example, the death rate on the *Merchantman* and the *Maidstone* sailing to colonial Guyana in the 1856-57 emigration season was 31.17 per cent and 24.53 per cent respectively. The complaint that the death rate tended to be higher and to increase in direct proportion to the numbers of women, children and infants on board was disputed by emigration officials who provided the quantitative analysis for 1850-58 (see Table 8.5).

Table 8.4: Percentage of Females on Emigrant Ships to Colonial Guyana, 1867-1868.

Year	Ship	Men	Women	Boys	Girls	Infants	% of Females
1867	Lincelles	200	102	31	16	8	33.8
	Indus	247	128	32	19	8	34.5
	Janet	230	198	40	24	11	45.1
	Cowan Clarence	261	128	27	19	9	33.8
	Oasis	241	92	23	12	7	28.3
	Orient	267	80	15	11	11	24.3
	Trevalyan	259	110	9	5	5	30.0
	Jason	291	92	11	4	7	24.1
1868	Ganges	307	88	7	3	5	22.5
	Clarence	369	73	10	7	5	17.4
	Adamant	269	60	16	7	5	19.0
	Harkaway	164	35	2	6	5	19.8
	India	249	85	19	6	13	25.3
	Howrah	298	97	18	4	16	24.2
	Trevalyan	271	90	16	15	15	26.8
	Winchester	297	103	27	8	18	25.5
	Himalaya	258	106	31	7	25	28.1

Source: CO 386, Letter Books of the Colonial land and Emigration commission.

Unacceptably high death rates would have given anti-emigration forces more ammunition with which to continue their opposition to a system of labour reallocation which, while not like the African Middle Passage in all respects, resembled it closely in many, leading Joseph Beaumont and Hugh Tinker to characterize Indian immigration and indentureship as the new slavery or a "new system of slavery" respectively. The "opposing voices" in Look Lai's terms or the "critics of indenture" in Mangru's, using arguments based on mortality, cost, treatment and so on, could already be heard from 1838 in colonial Guyana. Indeed, the earliest opposition was over the extreme cases of poor treatment meted out to some of the immigrants who were among the "Gladstone Experiment" batch. Opposition continued in colonial Guyana, as it did in Jamaica and Trinidad, over the entire period of the trade, from the Anti-Slavery Society and other groups such as the missionaries, the coloured middle class and the African-Caribbean freed people objecting to the scheme on the basis of cost, implications for freed peoples' employment and wages, and humanitarian concerns. There was also opposition in India over the suspicion that Indian women were being exported for the purposes of prostitution. This opposition later fuelled the female arm of the anti-emigration lobby in India in the late nineteenth and early twentieth centuries.

Table 8.5: Proportion of Women and Children Embarked on Ships from Calcutta to the West Indies, 1850-1858, with Per Cent Mortality

Year	Women to Men (proportion)	Children/infants to adults (proportion)	% mortality on whole no, embarked
1850-1851	9.09	5.11	3.61
1851-1852	16.93	10.89	4.45
1852-1853	23-96	16.53	5.60
1853-1854	14.36	7.84	3.30
1854-1855	18.34	7.48	2.75
1855-1856	35.72	10.82	5.75
1856-1857	35.27	14.67	17.26
1857-1858[a]	66.48	29.08	9.10

Source: CO 386/91, T.W.C. Murdoch and Frederic Rogers to Herman Merivale, 11 August 1858.
[a] Figures for *Salsette* of 1858 excluded.

In an effort to increase the numbers, of women available for emigration, some of the restrictions on female emigration were gradually relaxed, but the sexual disparity was never eliminated, though the records indicate that a great effort was made to conform to the ratios set after 1860. For example, the female to male ratio on the *St Kilda* to colonial Guyana in 1871 was 57.8:100. The ratio never fell below 40:100 on die ships which sailed for that territory in 1872, with this allowable ratio also being maintained during 1873.

Several incentives were given to recruiters in India in an effort to conform to the quotas set. The most widespread was the payment of higher rates of commission on every female recruited. For Suriname, recruiters were paid 25 rupees for each male recruited but 35 rupees for each female. In the nineteenth century, the rates for the British-colonized Caribbean territories were 45 rupees for males and 55 rupees for females. By 1915, as opposition to emigration grew and the supply became scarce, recruiting rates escalated to 60 rupees for each man and as much as 100 rupees for each woman. The rate for boys aged twelve to sixteen was half that of adult males. No incentives were paid for the recruitment of boys below the age of twelve, but the full female rate was applied to girls over ten and 20 rupees were paid for girls under the age often. This gender disparity, even among children, was due to the enormous need to increase the numbers of females shipped. There were charges that the payment of high commission rates, especially during times of extreme scarcity of prospective female emigrants, created the conditions under which kidnapping flourished. By the early 1870s, kidnapping had reportedly become prevalent in recruiting districts of Allahabad and the north-west provinces. The news paper, the *Pioneer of India,* carried a report in 1871 about an attempted kidnapping of an Indian woman in Allahabad by four peons, Gohree, Baldeo, Raoti and Rumzan. The woman was to be sent to Jamaica. The men were convicted and sentenced to prison terms ranging from six to twelve months.

There were charges of fraudulent behaviour in other aspects of the recruiting process. For example, planters exhibited a preference for married women, so recruiters, unable to fulfil this request to the extent required by the planters, manufactured couples for emigration. Admittedly, some of these alliances formed at the depots were consensual and once declared, all were accepted. The extent of the phenomenon of depot marriages is indicated by the returns contained in the protector of immigrants' annual reports. On the ship *Silhet,* which arrived in Guyana in 1883, thirteen depot marriages were registered. The numbers were at times higher, for example, in the same year forty-three marriages were registered on the *Berar* and forty-four on the *Bann*. But, as Brij Lal has observed, the number of married couples recruited in the normal process was higher than "depot marriages", and the phenomenon of depot marriages has often been exaggerated.

It was not unheard of for men and women to change their minds, and their partners, once on board or upon arrival in the Caribbean. For example, on the ship *Rohilla*, the agent general's report dated 24 March 1883 indicated that "the Nepalese woman, Moti, refused to acknowledge Amirbur [?] as her husband who is apparently an inhabitant of the plains. This couple was accordingly not registered as man and wife. On the ship *Foyle*, which arrived in Guyana in 1886, Asserum requested that she be located on a different plantation from Aladin, her "depot husband", as she had changed her mind about wishing to live with him. Indeed, she seemed to have made this decision on the voyage itself, benefiting from his protection on board, but never sleeping with him.

The castes from which the women originated continued to be a matter of concern to the planter class, which evinced a preference for higher-caste women. According to Mangru, the women who boarded emigrant ships for colonial Guyana comprised young widows and married or single women who had severed ties of relationships in India. He also claims that higher-caste women tended to be available for emigration in times of economic hardships in their villages.

Preference was expressed for young, healthy labourers. Planters requested young women in the age range of twenty to thirty except where they emigrated as a part of a family unit, in which case some age flexibility was allowed. The minimum age for girls to be recruited independently of a family was sixteen. In reality, recruiters often misrepresented die ages of die recruits, at times putting quite old people (with grey hair dyed black to disguise this) on board the ships. Such fraud was usually detected halfway through the journey.

The imbalance in the female to male ratio was initially reflected in the settler population being 11:100 in Guyana in 1851. Settlement in the colony at die end of indenture, increased importation and natural increase improved the ratio by the 1850s. The female to male ratio was 62:100 in 1858, 58:100 in 1891 and 73:100 by 1914. From a high of 61:100 in 1900, the ratio among rhe Indian immigrant population in Trinidad dropped to 42:100 in 1905 and 40:100 in 1914. In Jamaica the imbalance continued up to 1921, reaching 49:100 by 1943 (See Table 8.7) The male dominance continued in most importing countries up to 1946, not balancing out until the 1960s.

Women were recruited or captured from roughly the same geographical areas as men, though, as was the case under slavery, many were obtained far from their own villages, with the result that their area of origin was often wrongly represented. In the first phase of Indian labour migration, they were recruited from among those described as "hill people", from the Dhangar ethnic group from the Chota Nagpur division of die Bengal Presidency. Some originated among the poor elements of the cities of Calcutta and Madras and among the "untouchables" in the districts surrounding Madras. Some recruits or captives also came from the north-western regions and had been driven into the cities by famine in Upper India. Although some "tribal" people, including *santals* ("tribal" people from the Chota Nagpur Plateau of the Bengal Presidency), continued to arrive

Table 8.6: Percentage of Females on Emigrant Ships to Trinidad, 1863—1872.

Year	Ship	Men	Women	Boys	Girls	Infants	Total	%of Females
1863	Alnwick Casde	381	61	10	6	2	460	14.6
	Assaye	223	52	8	1	5	289	18.7
	Athlete	258	64	9	7	10	348	21.0
	Brechin Casde	180	71	17	11	8	287	29.4
	Golden City	388	56	7	5	3	459	13.4
1864	Alnwick Casde	339	83	15	13	7	457	21.3
	Spitfire	353	104	29	10	21	517	23.0
1865	Sydenham	307	79	15	19	4	424	23.3
	Atalanta	333	92	25	15	22	487	23.0
	Newcastle	403	79	19	17	20	538	18.5
	Carleron	297	97	48	37	14	493	26.4
	Empress	295	116	48	31	24	514	27.3
	Roxboro Casde	286	93	26	16	7	428	26.0
1866	Salisbury	258	127	28	22	10	445	34.3

1867	Alnwick Casde	233	166	30	32	17	478	43.0
	Sevilla	194	77	13	16	14	314	31.0
	Liverpool	329	159	22	13	9	532	33.0
	Ellenboro	195	136	14	9	6	360	41.0
1868	Sevilla	190	89	4	8	5	296	35.3
	Ancilla	243	68	8	3	5	327	22.0
	Malabar	330	82	12	4	7	435	20.1
1869	Poonah	217	116	22	12	20	387	35.0
	Arima	200	67	19	15	12	313	26.4
	Sevilla	163	80	26	25	19	313	35.7
	Ancilla	169	80	34	24	17	324	34.0
	Flying Foam	249	139	35	27	26	476	37.0
	Braurnaris Castle	192	142	73	54	43	504	43.0
	Wiltshire	217	210	77	56	-57	617	48.0
1870	Atalanta	188	118	42	23	26	397	38.0
	Cochin	288	120	35	16	23	482	30.0
	Java	279	124	28	24	20	475	32.5
	Hougomont	204	99	13	14	9	339	34.2
1871	Brechin Casde	274	110	19	17	12	432	30.2
	Indus	266	74	10	14	10	374	24.2
	Atalanta	275	81	9	9	7	381	24.1
	Syria	249	119	25	11	14	418	32.2
	Ganges	235	112	17	10	11	385	32.6
1872	Indus	250	104	20	16	23	413	31-0
	Woodburn	331	132	49	38	27	577	31.0
	Rajah of Cochin	218	111	44	26	14	413	34.3
	Delharrie	241	154	48	36	37	516	40.0

Source: CO 386/99, Colonial Land and Emigration Commission.

Table 8.7: The Asian Indian Population in Jamaica, 1871-1921.

Year	Male	Female	Total	% of Females
1871	5,339	2,454	7,793	31.5
1881	6,941	4,075	11,016	37.0
1891	6,338	4,467	10,805	41.3
1911	9,928	7,452	17,380	43.0
1921	10,203	8,407	18,610	45.2

Source: Jamaican censuses, 1871-1921.

in the Caribbean up to the 1860s, the main recruiting/captive areas by the late nineteenth century had shifted farther westwards as recruiters sought to combat the competition of the Assam tea gardens for workers from among the Dhangars and as planters demanded recruits who were not as subject to high mortality on the ships and plantations. By the 1870s major recruiting areas were the districts in the north-west Indian provinces of Orissa, the Punjab and Rajputana, and the various districts in the United Provinces of Bihar and Oudh. About 86 per cent of those recruited for colonial Guyana came from among the Bhojpuri—Hindi-speaking United Provinces. Only a minority were recruited from the Southern Madras Presidency from among both the Tamil-speaking and Telugu-speaking districts.

At times the percentage despatched to colonial Guyana from specific catchment areas varied from year to year depending on local happenings. For example, some participants in the so-called Sepoy Mutiny of 1857 were shipped to the colonial Caribbean. Look Lai notes that in the 1883—84 season, semi-famine in Bengal and Bihar caused recruits from these areas to be higher than usual among the emigrants. Recruits from these areas exceeded rhose from the United Provinces in the 1884-85 seasons and in the last nine months of 1885, the year in which the *Allanshaw* sailed. In general, however, the majority came from the United Provinces after the late nineteenth century. Raymond Smith concluded in a 1959 study that between 1865 and 1917, the year Indian labour migration ended, 70.3 per cent of the recruits to colonial Guyana came from the United Provinces and 15.3 per cent from Bihar, with a minority from other areas. Look Lai confirms the pattern further, stating that in 1898 a full 83 per cent of the 3,450 destined for Guyana from Calcutta came from the United Provinces. The female recruits or captives came from the same cross-section of religious and caste groupings as men, with lower Hindu castes predominating over high-caste Hindus and Muslims.

After being recruited or captured, both males and females, accompanied by *chaprasis* (messengers or orderlies), were taken to the depots in the area of first recruitment then transported by train to final embarkation depots either in Calcutta or in Madras. There were about ten Emigration Agencies serving the various colonies individually or in combination in Calcutta in the 1880s, but only three to four depots: the Mauritius depot at Bhowanipur, the Demerara depot at Garden Reach and others at Ballygun and Chitpur. Each was staffed by an Emigration Agent, doctors, clerks, watchmen and sweepers. These depots, all built on the same pattern, were not in the healthiest of spots, with grave implications for the health of those accommodated in them. They were surrounded by a high wall to prevent uncontrolled movement in and out and contained a number of barracks with bungalows for the staff. Accommodation was sufficient to provide for two shiploads at any time in the larger depots. Emigrants spent on average one to three weeks waiting to be shipped, although in extreme cases they could wait up to three months. On arrival at the depots, each recruit was told to bathe and issued with a change of clothing while the old clothes were washed and returned.

Since the process by which Indians were obtained for transportation was dogged by fraud, attempts were made by the various emigration officials to ensure that emigrants had voluntarily signed up for emigration. Despite the improvements and efforts to tighten control over the process, fraud and deception were never completely eliminated. Many recruits found themselves on ships bound for places they had not opted to go to or on journeys longer than they had been led to believe. Such fraud at times led to what emigration officials liked to style "mutiny". For example, there was a reported case of mutiny by emigrants on the ship *Clasmerden* from Calcutta to Guyana in 1862. The ship was forced to stop at Pernambuco in Brazil. The emigrants revolted because they claimed that they had been misled about the length of the voyage. Interestingly enough, some of those involved in this so-called mutiny had also been involved in the Sepoy Mutiny of 1857 and had been shipped out to the Caribbean, presumably to rid India of such "rebels". "Mutinies" were not confined to Indian emigrants. Chinese emigrants on the *Pride of the Ganges* bound for Guyana in 1865-66 rebelled over inadequate rations. Some of the emigrants who arrived in Trinidad in 1857 complained that Trinidad had not been their choice of destination; they had been led to believe that they were going to Guyana. These may have been encouraged to embark on a ship that was ready to sail, the ship for Guyana not yet being filled.

At the embarkation depots, emigrants were given several medical examinations, by the native doctor and the government-appointed European doctor, to ensure that they were fit to undertake the long journey to the Caribbean. The women were not given as detailed an examination as the men for two reasons: first, officials wanted few obstacles in the way of female emigration and opted for a cursory examination, and second, for cultural reasons, the detailed examination of women by male doctors was unacceptable to the women and the men, especially husbands. By the early twentieth century, agitation had increased for female doctors to be appointed to the depot to examine the women. Part of the preparations included administering the appropriate vaccinations and allowing the recruits to sign their indenture contracts. The emigrants were also issued with clothes for the journey: a pair of woollen trousers, jacket, shoes and a cap for males, and for females, a sari, two flannel jackets, one woollen petticoat, a pair of worsted stockings and shoes. With the exception of the sari, these clothes were not those of the typical rural Indian. In response to complaints about inadequate clothing before the 1860s, calls were made in 1867, for an additional petticoat to be issued to the women.

Once the emigrant was certified as fit for agricultural labour and embarkation, an emigration certificate was signed to that effect for each one by the surgeon-superintendent and the depot surgeon, and countersigned by the protector of emigrants at the port of embarkation and the emigration agent for Guyana. This certificate, duly signed and countersigned, was delivered to each emigrant. It contained the emigrant's name, father's name (Pargas[?] in the case of Maharani), age, caste, height, name of next of kin, marital status and name of partner, if applicable, distinguishing marks, and place of origin In India. Each emigrant also had a "tin ticket" or identification disk placed around the neck or arm, and an embarkation number. These details have allowed many descendants to trace their immigrant ancestors.

Once all the preparations were made, the emigrants were put on board the ships for their voyage to the Caribbean. The typical batch of women would have comprised a mixture of single, independent women who emigrated voluntarily to be involved in commodity production on Caribbean plantations; those who had been kidnapped or otherwise forced into emigration by depressed financial circumstances linked to the impact of British economic policies on the textile industry in India; those deserted by husbands and who were seeking a new life in a new country; a few (like the women Moorti and Mohadaya on the *Allanshaw*) who had previously been indentured in Natal, Mauritius, Fiji or the Caribbean, and who had opted for return and reindentureship; wives and dependents accompanying husbands or other relatives; and those (such as Chandra Kumari, who in 1891 embarked on the ship *Erne* for Jamaica with her boyfriend, Tek Bahadur) who were simply out for a new adventure. Chandra was at first located on the Belvedere Estate in Jamaica but refused to do any work. She claimed that she was the daughter of the king of Nepal and that she wished to be released from indenture. Under Jamaican Law 23 of 1879, her indenture was cancelled. Further research into the matter revealed that she was not "royalty" as she had claimed. The protector of immigrants was told by Mr Wylie who did the background check in India that Kumari "was either a runaway slave girl from one of the nobles' establishments in Nepal or a courtezan who has gone off with Tek Bahadur". Charles Doorly, the protector of emigrants at the Madras agency, added as factors of emigration, "domestic unhappiness caused by quarrels with their husband's relatives or with the other wives; widowhood and all its attendant miseries".

THE SHIPS: SPACES OF (S) EXPLOITATION?

Recruits were transported from Calcutta (as were those from Madras) in ships specially selected to participate in the Indian labour migration scheme. They were also placed under the care of crew members who were instructed in their responsibilities towards those in their care. A primary concern of those who decided on the ships to be used and the crew to be hired was the necessity of

Figure 8.1: The *Allanshaw*, 1874 (*Illustrated London News*, 12 November 1881, 472)

avoiding any charge that Indian emigration was a new system of slavery. As the way in which a society treats its women is considered a measure of "civilization", and since the condition of African women on the Middle Passage and on the plantations had featured so highly in the anti-slavery campaigns of the late eighteenth and early nineteenth centuries, emigration officials were particularly concerned that the regulations concerning the treatment and care of Indian women on emigrant ships should be observed. This section of the chapter outlines the major regulations and explores the extent to which emigrant ships were efficient transporters of labourers or simply "large spaces of (s)exploitation".

REGULATIONS FOR EMIGRANT SHIPS: MANAGING SPACE AND HUMAN CARGO

Specific standards relating to size, space, crew, diet and medical stores were laid down for ships accepted to participate in the Indian labour migration scheme. In general, the ships were larger in tonnage than those used in the transatlantic trade in enslaved Africans. Indeed, Herbert S. Klein records that between 1782 and 1788, the size of ships engaged in transporting enslaved Africans to Jamaica averaged just 172 tons. The average size increased to 236 tons between 1791 and

1799 and 294 tons between 1800 and 1808. The average number of captives carried per ship in these same periods was 396, 328 and 289, respectively. The Parliamentary Slave Trade Regulating Acts of 1788 and 1799 had reduced the number of captives per ton ratio as a way of reducing the overcrowding and high mortality on slavers, because in order to compensate for inevitable deaths, slavers had been packed way over their capacity. After 1799, the regulations provided for one captive per ton as opposed to the previous 2.6 per ton. The tonnage of ships also rose after 1788, though few ever approximated the size of the emigrant ships. The emigrant ship *Whitby* to colonial Guyana in 1838 was among the smallest ships used in the trade in Indian contract labourers, being just 350 tons. By the 1850s, a larger class of ships was being used. Some, like the *Ganges* owned by James Nourse, were 839 tons. The Nourse Line became the principal carrier to the Caribbean, although Sandbach, Tinne and Company was also involved. From the 1880s, ships increased in size to 1,600 tons or larger; but smaller ships were not eliminated, the *Salsette* which sailed from India to Trinidad in 1858 being 579 tons. Between 1884 and 1888 Nourse brought seven iron ships into service, including the 1,674 tons *Allanshaw* (built in 1874 by Renfrew, W. Simons and Company and registered at Lloyds of London). Steamships made the voyage in less time. For example, the first sream ship to Guyana, the *Enmore*, took just 49 days. The journey on sailing ships, like the *Allanshaw*, could take up to three months, sometimes longer. The *Silhet* took 96 days to reach Jamaica in 1878 and the *Lightning* 112 days, while the *Salsettes* journey lasted 108 days. Despite their longer sailing time, sailing ships were never totally replaced, and in 1895 there were still twenty-two of them in service. Slavers from West Africa had taken four to six weeks on average. Though some ships passed through the Suez Canal, the usual route, such as the one made by the *Allanshaw*, was around the Cape of Good Hope.

Ships had to be well ventilated and, from the 1860s, were required to be supplied with lifeboats, fire appliances and other rescue equipment. No firearms or flammable materials were allowed on board. Emigrant ships had to carry an adequate number of officers (referred to as "mates" on merchant ships such as those in the Indian labour trade), crew and deckhands. The personnel usually consisted of the captain, who was in overall control of the ship and supervised the officers and crew, the surgeon-superintendent who had responsibility for the "passengers" or emigrants; three officers, each with special responsibilities (for example, *the* third officer was also the boatswain); and the crew comprising ordinary sailors, "able(-bodied) seamen", deckhands and workers with special skills (some drawn from among the emigrants), such as cooks, sweepers (usually called a "topaz" or "topass"), nurses, hospital attendants, and dispensers/compounders. Respected men, usually second-time emigrants, drawn from among the Indians, were designated "sirdars" or leaders, and assigned to supervise the emigrants in a ratio of 1:25.

Once the ship had embarked all its "passengers", the officers and crew were mustered on the quarterdeck and reminded by the captain of their duties and responsibilities on the voyage. Failure to carry out assigned duties to the satisfaction of the captain and surgeon-superintendent could result in loss or reduction of gratuities and even jeopardize future employment on emigrant ships. This could have affected the statements given by those called to give evidence before the Commission of Enquiry in colonial Guyana after the voyage of the *Allanshaw*. In the special case of the Indian labour trade, even captains and surgeons-superintendent could lose future employment on emigrant ships if investigations found them negligent in their duties.

The ship's surgeon-superintendent was crucial to the voyage and great effort was made to select competent and experienced ones. As Mangru observed, the rate of mortality among the emigrants depended considerably on the care, competence and character of the surgeons-superintendent employed. The positive correlation between competent surgeons and low mortality on some ships did not escape the attention of emigration officials. Despite the acceptance among all involved that the role of the surgeon-superintendent was essential to the success of the Indian labour migration scheme and that the best qualified ones were to be selected, the surgeons-superintendent were inevitably a mixed group of competent and not-so-competent men. Some of them had a wealth of experience in the Australian service, some were former army doctors, and others were inexperienced men employed in the Indian medical service and Indian gaols. As was the case with the ships' doctors under slavery, not many qualified surgeons wanted to participate in this "disagreeable trade". Falconbridge, in discussing the surgeons on slavers, observed: "It may not be improper here to remark that the surgeons employed in the Guinea trade, are generally driven to engage in so disagreeable an employ by the confined state of their finance." Strained finances might have been a major reason for the surgeons-superintendent to join emigrant ships, though their salaries were far from attractive. Indeed, the low remuneration may have been one factor militating against the ability of emigration officials to secure more competent surgeons. Surgeons-superintendent received a gratuity per head on those landed alive with an increase in proportion to the number of voyages undertaken. A surgeon-superintendent such as Hardwicke on the *Allanshaw* would have received a higher rate of gratuity than some others as he had undertaken about nine voyages in this service. This per capita pay was lower than that paid to those transporting emigrants to Australia; consequently, the better ones gravitated towards the Australian service. Those on the emigrant ships were guaranteed a free return passage to India or England, but they were not entitled to a pension, free outfits or continued employment in the service. Besides gratuities, surgeons proceeding to the Cape of Good Hope or Australia were given fixed sums of £40 and £60 respectively, claimable within a specified period, in lieu of return passage.

The continued difficulty of attracting competent surgeons resulted in an increase in their pay by the late 1860s when it went up from 8 shillings a head landed alive to 10 shillings for the first voyage, 11 shillings for the second, as long as the surgeon's conduct was deemed satisfactory, and 12 shillings for subsequent voyages. Still, a veteran in the Australian service could earn up to 20 shillings (£1) a head. The Emigration Act XLVI of 1860 also stipulated the appointment of European surgeons or Indians who had received collegiate training in the European system of medicine (though the lack of respect accorded Indian surgeons on board the ship made them less likely to join an emigrant ship). The nationality and skill requirements were not always easy to meet, and thus were not always observed when prospective surgeons were scarce. Where the European surgeon was not fluent in the language of the emigrants, an interpreter was provided. At the same time, Act XIII of 1864 laid down in greater detail the duties of the surgeon, which started before the ship departed the port of Calcutta or Madras. Both surgeon and compounder were to be appointed at least ten days before the ship was scheduled to sail. The surgeon was to visit the depot daily, or at least five or six times, to examine the emigrants individually, inspect the ship's ventilation, hospital, privies, water and distilling apparatus, and report any defects or deficiencies. However, the surgeon was never given the ultimate responsibility of deciding on the fitness of emigrants to embark. There were continual complaints from the surgeons, especially when mortality had been exceptionally high on a particular voyage, that many who embarked should never have been put on board by the emigration agents.

On the voyage the surgeons-superintendent were required to attend to the medical care of the emigrants on board the ship, making sure to detect and treat illnesses before they assumed epidemic proportions. They were to "keep up spirits" of the emigrants by encouraging them to sing and dance, and supervise the cooking and serving of food. Because of the difficulty of attracting competent surgeons to the Indian labour migration trade, infractions of the rules laid down for their conduct were often overlooked; still, there were cases, however infrequent, of withholding of gratuities and outright dismissal. Some carried out their duties diligently, others did not.

Up to the 1850s males and females had been separated on board without 'any attention to family, but this was later modified. The strict spatial separation by gender had been maintained on the grounds of "decency" and to reduce the spread of sexually transmitted diseases by limiting, or trying to prevent, sexual contact. It was the negatives of this arrangement which dictated later modifications. The negatives stressed were that families and couples were separated and that there was an increased likelihood of sexual harassment of women by sailors. Indian men, the supporters of modification argued, were necessary to protect Indian wives on board. Consequently, by the 1860s single men and women were separated but couples were kept together. Single women were placed aft (the rear of the ship), married couples and children were amidship, and single men were placed in the forward part of the ship.

Regulations were laid down regarding clothes and diet, the space allocated to each emigrant and the sleeping arrangements. The food supplies provided for Indians conformed to caste and religious preferences. By the 1870s, meat (mostly fresh mutton) had been introduced into the diet on board instead of dried fish which had previously been the choice. Dried fish was still recommended for the first few weeks of the voyage when unstable sea conditions made cooking difficult, but fresh meat was to be cooked thereafter. Increased portions of rice, flour, ghee and dhal were implemented in 1871. The cook was usually drawn from among the highest Hindu castes on board in an effort to avoid caste-conflicts, but such conflicts could not be avoided completely. Furthermore, the choice of Hindu cooks (and, at times, what they chose to serve) offended Muslims, usually a minority on emigrant ships. Twenty Muslims on the *Jura* bound for Guyana in 1891 reportedly refused to eat because the sheep had been killed by a Hindu. The food was thrown overboard. On arrival of the ship *Grecian* in Georgetown in 1893, the agent general of immigration reported that three Muslims had been placed in irons on the voyage on 17 September 1892 for inciting the other immigrants not to eat the food issued to them because the meat was pig and beef not fit for Mohammedans. Of course, the objection might have been related more to the fact that the meat was not *halal* (prepared as prescribed by Muslim law) than because of the ethnicity of those who prepared it.

The use of space on emigrant ships was regulated as a result of the concern about avoiding the overcrowding and high mortality so much associated with slavers in the African Middle Passage. In 1842, the numbers of migrants allowed on board were related to the size (tonnage) of the ships, one emigrant to two tons. By 1845 this was abandoned in favour of the cubic measurement of seventy square feet or twelve "superficial" feet for each emigrant. The colonial land and emigration commissioners were insistent that sufficient space should be allocated to allow for "respiration and motion" and that such space should be such that a "full grown man" of about an average height of five foot, six inches "might lie down, supposing no system of berthing [on-ship beds or bunks] be adopted [and] move about without much inconvenience". Despite the opposition from some quarters, Act XIII of 1864 reduced this to ten superficial feer for each adult. Increased numbers could be carried by this reduction. This allotment again caused overcrowding and inadequate ventilation, and so was reversed by the Colonial Land and Emigration Commission to the 1845 space for adults. The allocation often superficial feet was retained for children. No upward limit of numbers was actually stipulated, but preference was shown for larger vessels with better ventilation and space for recreational facilities than smaller vessels. There had been an attempt to stipulate that emigrant ships should not carry more than 300 to 350 contract labourers on board, but this had been abandoned in 1860 on the grounds that not only did it increase the cost of passage but it restricted the emigration agent's selection to small ships which could not be as well fitted out or ventilated for comfort as large ships.

Initially, *the* emigrants slept on mats on the deck of their quarters. In 1866 a recommendation was put forward by Dr Pearse, surgeon-superintendent on the *Oasis*, for raised wooden platforms instead of mats. This was in an effort to improve health and reduce the mortality on ships, but it is unclear if this recommendation was effected on all ships involved in the indentured labour trade.

REGULATIONS CONCERNING THE CARE OF THE EMIGRANTS

Strict rules regulated the conduct of the sailors towards the emigrants. For example, when the forty-six men were mustered on board as the *Allanshaw* started its journey, they were reminded that "any member of the crew found amongst the emigrants talking to, or interfering with, or molesting them in anyway, will be fined one month's pay each offence", and Robert Ipson seemed to have been reminded about this constantly by Captain Wilson and Dr. Hardwicke. Fraternizing between the crew and Indian women was especially forbidden. The hatchways were guarded, especially that section leading down to the single women's quarters. Men were forbidden to enter this section of the ship, and this applied to both fellow emigrants and crew, whose quarters vrere usually in the forecastle, next to the prow. Some surgeons ensured that die decks below were well lit, especially the female section, to prevent what they termed "promiscuous intercourse".

The strict legislation regarding protecting women from the crew and single male emigrants makes it clear that there was a real fear that emigrant women were always in danger of sexual assault. Indeed, Seenarine recently concluded that "the entire 'coolie ship' was an unsafe place for single females, as well as married women, as they were frequent targets of sexual attacks".

Regulations required that adequate food be provided for the emigrants. The sirdar received and distributed daily rations, helped to supervise the cooking and sanitary arrangements in addition to his other role of assisting in the maintenance of discipline and promoting the emigrants' general welfare. Emigrants were normally fed twice per day and the emigrants usually took their meals on deck as long as the weather was good. Breakfast was served by nine o'clock and the evening meal between five and six o'clock.

The emigrants were encouraged to be on deck as much as possible and to entertain themselves by dancing, singing and playing games. Adult men and women were provided with chillum pipes and women were provided with combs and other supplies for their comfort.

Every attention was supposed to be given to die health of those on board, particularly with a view to maintaining as low a death rate as possible. Emigrant ships were fitted out with water closets, separate ones for men, women and the crew by 1885. This not only prevented their quarters from being smelly from excrement but also reduced the spread of diseases on account of a filthy environment. Emigrants were encouraged to bathe at least once a week, to keep active and to oil themselves with coconut oil weekly. After the near disastrous 1856-57 emigration season in which the death rate on ships from Calcutta to the Caribbean ranged from 6 per cent on the *Wellesley* to 31 per cent on the *Merchantman,* an average of 17.27 per cent, regulations were improved with a positive impact on the death rates which had fallen to 3 per cent by the 1862-63 season. Apart from enormously high death rates when accidents happened - as in 1858 when 120 out of 324 emigrants on the *Salsette* to Trinidad died en route, when in 1865 the *Fusilier* was wrecked at Natal with a total

Table 8.8: Mortality at Sea: Voyages to Colonial Guyana, 1871-1890.

Year	%	Year	%
1871	1.60	1881	2.68
1872	4.74	1882	1.46
1873	5.56	1883	0.64
1874	5.58	1884	2.04
1875	1.12	1885	2.50
1876	1.08	1886	1.41
1877	1.52	1887	1.59
1878	3.30	1888	1.82
1879	1.55	1889	1.50
1880	1.34	1890	1.41

Source: D.W.D. Comins, *Note on Emigration.., to Guyana, 1893,* in Hugh Tinker, *A New System of Slavery: The Export of Indian Labourers Overseas* (London: Oxford University Press, 1974), 165.

loss of life of 246 (including those who died from fever on the voyage to Natal), and when the *Eagle Speed* was wrecked at the mouth of the Mutlah River with 300 lives lost—in general, by 1885 the mortality rate was much lower. It was 3.26 per cent on the *Jorawur* which landed in Guyana in December *1884* and under 3 per cent on the *Allanshaw* of 1885. The mortality rate was lower on ships sailing from Madras (an average 0.9 per cent in the 1850s) than on those which sailed from Calcutta. Among the improvements were better medical examination and selection of prospective recruits at the emigration depot in India, improved ventilation, separate compartments for the sick, recruitment of experienced surgeons capable of dealing with diseases and illnesses on board, improved water, and better diet on board. By the time the *Allansbaw* sailed, the average mortality was 2.50 per cent on ships to colonial Guyana, and it was to become even lower by 1890, as shown in Table 8.8.

Despite the strict regulations governing the treatment of emigrants on board ships to colonial Guyana and other importing territories, such rules were flouted constantly and flagrantly. In 1862, for example, charges were brought against the surgeon-superintendent of the *Persia*, Mr. Chapman, for cruelty to the emigrants on board. In 1863 those on the *Clasmerden* who had staged the so-called mutiny complained that the surgeon, A.N. Watts, had been drunk for roost of the voyage and had not carried out his duties satisfactorily. His gratuity was eventually reduced as punishment from 10 shillings to 2 shillings per emigrant landed alive. On the *Jorawur* 1884 complaints of indiscipline, lack of authority and drunkenness on the part of the captain were made. There were complaints on several voyages that the food was inadequate, that potato, milk and pumpkin spoiled on the way, and that food designated for the emigrants was at times siphoned off for the crew. There were also reports that the water taken on board was contaminated and that emigrants were physically punished by the crew. On the voyage of the *Main* to Guyana in 1888, several crew members breached the rules proscribing physical punishment of emigrants and proceeded to shackle and handcuff those they regarded as having committed offences. Whereas African men were routinely shackled and handcuffed on the Middle Passage slavers, regulations surrounding Indian emigration were that emigrants were to be put in irons only in extreme cases of indiscipline, and only the captain could do this. But the interpretation of what constituted "extreme indiscipline" was often suspect. Putting emigrants in irons was also allowed if they threatened to jump overboard to commit suicide (as several high-caste Indians had done on the *Foyle* of 1887, following the example of Podrath Singh, because they claimed to have been "polluted" by being touched by a low-caste sirdar), or if they were deemed insane. The eight men and five women who had been handcuffed on the *Main* complained to the agent general of immigration upon arrival in Guyana that they thought the punishment had not fit the so-called crimes. One of the women had been handcuffed, for example, for lighting her *haka* (chillum pipe) and smoking in between decks.

Despite the insistence of the emigrants on the *Allansbaw* that they had been well treated on the voyage (see chapter 3), that journey had not been incident-free. The logbooks kept by the surgeon-superintendent and captain were full of incidents in which the crew abused the emigrants. For example, the "coloured boatswain" had struck Auntoo on the shoulder with a rope; Juggessar and his wife had been verbally abused and threatened by Robert Ipson; Ipson had abused others, including the cook and "KaluVKaloo no. 289" whom he had pushed and kicked on 30 September. This led Kalu to say to two of his compatriots, Janki and Bhadaya (interpreted for Ipson by a sailor, Templeton), that Ipson might have been one of those who had raped the young woman, Maharani; Nandhal had been struck with a rope by the engineer; Palukdhan had been struck by the sailor John Smith; O'Brien had deliberately thrown a six-pound tin of mutton at the Indian women on deck; Beharie's thumb had been cut by the cabin boy William Clintworth who was subsequently punished, leading to a near mutiny by the crew; and several sailors had been accused of annoying the hospital inmates "when they came aft to muster at 8 p.m. [on 23 August] by making ugly noises with their mouths at the doors and windows".

The regulations against fraternizing between crew and female emigrants and proscribing sexual relationships between crew and Indian women were also flouted constantly. If the ships' records and immigration reports are to be believed, the sexual exploitation of women in this period did not reflect only white but also black and Indian masculinity in action, and there *is* proof that the emigrant ships hired non-European crew, many of whom were black. Indeed, there were at least six black sailors on the *Allanshaw*. Black men (African, Caribbean and US citizens) had a long history of maritime occupations, a fact that is not often recognized. But as W.

Jeffrey Bolster recently recorded, the stereotypical view that "blacks aboard ships sailed as commodities rather than seamen" needs to be overturned. He adds that even before the abolition of slavery, individual enslaved men routinely drew on maritime work to take charge of their lives or to communicate with distant blacks: "free and enslaved black sailors established a visible presence in every North Atlantic seaport and plantation roadstead between 1740 and 1865". This was as true for US blacks as for African-Caribbeans. A post—War of Independence shipping boom in the United States created jobs for enslaved and freed African-American men, and since "seafaring in the age of sail remained a contemptible occupation for white men, characterized by a lack of personal independence and reliance on paltry wages", according to Bolster, it became an occupation of opportunity for black men. Shipboard work for blacks became less significant after the abolition of slavery in the United States when white men increasingly dominated this occupation, so many African-American seamen sought jobs with the English shipping companies. This helps to explain the presence of "Yankee sailors" (a description applied to Ipson who had served in the American navy) aboard nineteenth-century Indian emigrant ships. Indeed, apart from Ipson (from the Danish Caribbean), the *Allansbaw's* crew list made it clear that British maritime commerce hired seamen from countries outside the British Empire, including Brazil, Germany and Finland. It has already been noted that Indians, some passengers themselves, were also hired as crew. A few positions, like that of sirdars, were held by high-caste Indian men and a few cases have surfaced in the manuscript sources of these sirdars molesting women on board. Sweepers and cooks also tended to be Indians.

Race was never irrelevant aboard ships. As Bolster observes: "Black men understood that among sailors, race worked in an ambiguous and sometimes contradictory fashion." So, in general, while seafaring was a way of escaping the discrimination in post-slavery societies and facilitate upward social mobility, certain roles on the ships were assigned to blacks on facial grounds. On the Indian emigrant ships, while black men never held the lowliest positions, which seemed to have been reserved for subaltern Indian men for cultural reasons (for example, Indians would never agree to black men cooking their food), they tended more to be rank-and-file seamen rather than officers for the most part. But even as seamen, they held some position of authority over the bonded labourers on board. Still, while black men may have felt superior to their "culturally alien" "human cargo", they were keenly aware of the racial hierarchy *vis-à-vis* European crew. Bolster records that black men often suffered disproportionately the capricious nature of shipboard punishments and, I daresay, discipline. Despite collective work and an easy familiarity between non-European sailors and their white shipmates, social identities were still conditioned significantly by race, and many white seamen just did not like non-white or black seamen. Ipson's claim that he had been singled out for particular attention by the *Allanshaw's* officers might therefore not be unbelievable; neither might be his claim that some of the Indian emigrants deliberately lied about his involvement. For it is undeniable that inter-racial prejudices coloured inter-ethnic relations on board nineteenth-century Indian emigrant ships.

However, although there was no love lost between European and non-European, especially African and Caribbean, crew, as is indicated by the frequent outbreaks of conflicts between them, and the equally frequent requests by the surgeons-superintendent of the ships that black crew members be replaced by Europeans, it would seem that the actions of both groups towards Indian women reveal that indentureship, like enslavement, manifested signs of gendered tyranny on the part of those who exercised, and abused, their power over the emigrants they were exclusively employed to protect. Although the duty of the seamen and officers was to see to the health and comfort of the emigrants aboard the ships engaged in *the* trade in contract labourers, they often held out rights as "privileges" in order to force compliance among the women. Above all, the actions of black and white men towards Indian emigrant women on the ships destined for the Caribbean demonstrated that the roots of the racist and ethnic tensions which later characterized the host societies in which Indians settled, were deeply embedded in the voyage from India and did not suddenly emerge in the region. As settlers, both Indians and African-Carib-beans harboured mutual feelings of contempt and superiority towards one another; such feelings, no doubt, had already found expression on the passage from India. Thus Donald Wood's view that Indians "encountered Negroid peoples for the first time in their lives when they landed in Port-of-Spain" is not completely accurate. The whites on board, both officers and rank-and-file sailors, harboured a certain contempt for Indians as a race, perhaps a racial attitude towards non-whites inherited from enslavement and strengthened by British imperialism in India, and the emigrants as a class. One surgeon-superintendent, Dr. R.

Whitelaw, noted in his diary of 1882 that "there is a great tendency among officers, apprentices and men (if European) to consider the coolies [sic] a people who may be pushed about, abused and annoyed at will". Even Ipson articulated clearly that he did not know what all the fuss on the *Allanshaw* over the emigrants' welfare was about: he "had been in emigrant ships before and with as many as 800 on board and never saw such a damned fuss as these was made on board of this ship about them". Their views about lower-caste women were clearly seen in their treatment of female emigrants on the voyages and came out strongly in the evidence given at several official enquiries launched over the period of indentured labour migration when complaints about sexual assault and other misconduct of the crew could not go unnoticed or pushed under the mat. The rather "democratic" spread of the practice among the men in charge of the ships, as this section will demonstrate empirically, reflected a certain amount of acceptance of it.

True or not, the complaints about the abuse of emigrant women by black men were seen in the reports of emigration officials and ships' surgeons. Indeed, one argument used by the surgeons who called for the reduction in the number of black men employed as crew on Indian emigrant ships, or their total elimination, was the usual ethnic stereotype that they had an "incorrigible addictedness to sexual intercourse". This view was articulated strongly by the surgeon-superintendent of the *Moy* which landed in Jamaica in June 1891 and whose crew apparently consisted mostly of black men. He wrote in his report:

> the greatest difficulty was experienced during die voyage in preventing intercourse between the [black crew] and the female passengers. I concur in thinking that on account of their generally incorrigible addictedness to sexual inter-course, negroes [sic], if so employed, should be in a minority on a cooly [sic] emigrant ship.

Individual black men were implicated on several occasions, before and after the case on the *Allanshaw*. The Agent General of Immigration, in summarizing all the data relating to the voyage of the *Avon* to Guyana in 1892-93 for transmission to the Colonial Office, reported that it had come to his attention that Steed, an African crew member "had so often been cautioned against interfering with or molesting the immigrants that the surgeon-superintendent, on arrival at St. Helena requested the Inspector of Emigrants to have [him] removed from the ship". The inspector of emigrants refused to take this drastic action on the basis that no actual assault had taken place and so no legal charge could be brought against Steed. Although some surgeons-superintendent complained that black men were more likely than white men to molest women on board emigrant ships and used this argument to justify their call for a white only or predominantly white crew, a higher proportion of the complaints about the sexual abuse of emigrant women by officers and crew on emigrant ships to the Caribbean was directed at European men. The surgeons themselves were infrequently implicated, based on complaints made by the emigrants on arrival at their destination. My research has uncovered more evidence of "sexploitation" on the ships destined for Guyana than on those destined for other Caribbean territories. This may be a reflection of the uneven nature of the contents of the reports submitted by emigration officials and surgeons. It could also be explained by the fact that far more reports exist for Guyana which outpaced any other receiving territory in the region, and was the destination of numerous ships from Calcutta and Madras.

The major sources of evidence are enclosed with the correspondence between the emigration agents and the land and emigration commissioners for the period 1854-60 and with the correspondence between the colonial governors and the Colonial Office officials. It was customary for all reports of the surgeons and agents general of immigration/emigration (protector of immigrants/ emigrants in some territories) to be forwarded to the Colonial Office and these have proven to be a wealth of information on the issue of the abuse of women. The reports of the agents-general of immigration in colonial Guyana are replete with complaints of "misconduct" on the pan of the surgeons-superintendent, captains and crew, for immigrants were usually encouraged to lodge such complaints on arrival in the colony. Such complaints were made against Dr. Wilkinson on the *Bucephalus,* Dr. Galbraith on the *Devonshire,* Mr. Simmonds on the *Royal George* and Dr. Cook on the *Assaye* — all to Guyana. The captain of the *Thetis* was said to have indicated that he had no intention of interfering when the Indian women and the sailors on the ship engaged in sex, and the surgeon-superintendent of the *Canning* of 1860 lost a part of his gratuity because he was accused of getting three women drunk so that they could not testify against him. On that same ship, two sailors had "violently assaulted" women to have sex with them, yet no-one found out who the individuals were so that they could be punished. The crew on the *Dovercastle* to Guyana in 1871 was accused of "misconduct" towards the women on board. The land and emigration commissioners complained that as long as the water closets were placed where they were [in the fore of the ship] "these things will happen". Murdoch of the Colonial Land and Emigration Commission, in a letter to R.H. McCade, indicated further that "we have always maintained the impropriety of so placing them, notwithstanding the opposition for several years of the Indian Authorities". There were complaints that the drunken captain of the *Jorawur* of 1884 and the crew and the steward of the *Grecian* of 1885 molested the women.

Dr. Atkins, surgeon-superintendent on the ship *Silhet* to Guyana in 1882, was said to have formed a relationship with the female emigrant Janky despite the fact that Deemohammed, a sirdar;, had staked his claim to her previously. Predictably, Atkins was absolved of any "illicit intercourse" with Janky during the voyage and was later allowed to marry her, a highly unusual occurrence for the nineteenth century. On arrival in Guyana, Atkins requested the cancellation of Janky's indenture on payment of the required sum. After this was agreed, but before the final permission to marry was granted, a contract was drawn up "securing to the wife control over the sum of £250 deposited on her behalf by the husband with the Acting Administrator General, to be applied for the benefit of the said wife as he and his successors should think proper". The sum was invested with the receiver-general. Janky was given a letter addressed to Messrs W. and H. Brand of 109 Fenchurch Street in London, who were agents of the Department of the Administrator General. This letter instructed them to give Janky any advice she needed in the event of her husband›s absence or death. They were married in Georgetown, Guyana, by the minister of the St. Andrew's Church, and Atkins was directed to present himself to the emigration agent on arrival in London, where he proceeded to take Janky.

Such complaints about the surgeon, captain and other crew on emigrant ships were not confined to Guyana but applied to Jamaica and Trinidad as well. An early case to come to light on a voyage to Jamaica involved Dr. Prince, surgeon-superintendent of the *Ravenscraig* of 1861. In addition to having been accused of excessive drunkenness on the voyage, Prince was said to have committed "criminal assaults" on some of the Indian women "under circumstances of an extremely aggravated nature". As the alleged offences were committed before the Atlantic crossing and, as was usual in such cases, the ship stopped at St. Helena, the governor there ruled that there was sufficient evidence against Prince. He ordered that Prince be sent to Jamaica under arrest and relieved of his duties. On Prince's arrival in Jamaica, however, the police magistrates and other officials released him on the basis that there was not enough evidence to convict him. They paid him his salary and gratuities as set out in his contract. The police magistrate and his supporters felt that even if such intercourse had taken place between Prince and the women on board, "it is clear that it must have been with their consent". How they came to this conclusion is unclear, but it was not unusual for lower-caste women from India, as African women enslaved in the Americas had been, to be regarded as loose and promiscuous and incapable of being raped.

The reports of voyages to Trinidad also contained complaints about illicit intercourse between crew and emigrant women. Complaints were made in this regard about the *Nerbuddah* of 1885. Complaints also originated among emigrants to Mauritius and one Dr. R. Brown was actually dismissed from the Mauritius service after four voyages when he was reported for drunkenness and pulling off the clothes of female emigrants.

It should also be pointed out that Indian women also suffered abuses at the hands of Indian men on board as domestic violence was not unheard of in instances where people tried to live as couples and families on board. On the *Artist* to Guyana in 1874, there was "a murderous assault on a woman by her husband", who was subsequently "put in irons to be dealt with upon landing". "While on the journey of the *Silhet* to Guyana in 1883, the male emigrant Gazee and his wife had an argument. She complained, and the third mate hit Gazee and threatened him with further action if he continued to ill-treat his wife. The surgeon-superintendent's report indicated that Gazee seemed to have believed that he had "a prescriptive right" to ill-use his wife. Perhaps Gazee's attitude was reflective of the caste and gender hierarchy which were organizing principles of the Brahrninical social order and which resulted in the subordination of women.

A male emigrant on the *John Davie* in 1885 reportedly tried to rape a young girl. He was beaten, even though corporal punishment was not supposed to be carried out on emigrant ships. The captain and surgeon-superintendent, however, felt justified in breaking the regulation in view of the gravity of the offence. The man was given six lashes "with a moderately hard rope on the buttocks".

The protector of immigrants in Trinidad reported that a woman, Bhagwandie, a passenger on the *Nerbuddah* of 1885, had jumped overboard "after she had been assaulted by a man... who appears to have slept with her several nights during the voyage". It turned out that this man was Indian, a sirdar, who had also stabbed her in her side with a bayonet. He was arrested on arrival. Several Indian sirdars and Lascars were also accused of molesting Indian women on this voyage.

The foregoing examples should indicate that the sexual abuse that Maharani claimed to have experienced on the *Allanshaw* before its arrival in Guyana in 1885 was not implausible. Emigrant ships were not only transporters of contract labourers to colonial plantations, they were also "spaces of (s) exploitation". This should be borne in mind when analysing the findings of the Commission of Enquiry held to investigate Maharani's death. Also to be borne in mind is the racism at work on nineteenth-century emigrant ships which targeted black men for exposure far more than European men and Indian men, even though the latter were clearly not blameless in the "sexploitation" of Indian women.

CHAPTER NINE

Tourist-Oriented Prostitution in Barbados:

THE CASE OF THE BEACH BOY AND THE WHITE FEMALE TOURIST

Joan L. Phillips

This ethnographic study focuses on the relatively recent phenomenon of male tourist-oriented prostitution in Barbados. It explores issues of gender identification and negotiation between the white female tourist and the black, often dreadlocked "beach boy." Predominantly qualitative research methods were employed in order to gain an emic (insider) view of the phenomenon and discover how these actors conceptualized their situation.[1] The study argues that male tourist-oriented prostitution is based on a quest for the sexual Other. This quest is structured along racial and gendered lines, where the white emancipated Western female goes in search of the quintessential hypersexual black male in the center of the Other.

PROSTITUTION IN BARBADOS

Prostitution in Barbados, although deemed illegal by common law and more recently in the Sexual Offences Acts of 1992 and 1998, has had a long, informal yet covert subculture in Barbados. The predominant attitude of the state toward prostitution is based on the view that prostitution as an activity is detrimental to the moral fabric of society. With laws against loitering, brothel keeping, and soliciting, the Barbados government has made its stance quite clear. For example, the Minor

Offences Act of 1998 states: "Any person who loiters in any street or highway and importunes passengers for the purpose of prostitution... commits an offence and is liable on conviction before a magistrate to a penalty of B$2,500 or to imprisonment for 2 years or both."

However, that is not to say that prostitution is not a culturally accepted practice. The few studies that have focused on the issue of prostitution have looked at female prostitution and pointed to Barbados's colonial legacy in playing a part in its development (Kempadoo 1996). Beckles, focusing on the pre-emancipation period, 1650-1834, in Barbados argues that "slavery in some ways corrupted the sexual values of the inhabitants" (1989:141). He points to the fact that in Bridgetown organized prostitution and the keeping of resident mistresses was very much the norm. This practice was further facilitated by liberal urban values. The geopolitical significance of Barbados as an important seaport provided a steady demand for sex. Levy argues that "black prostitution during slavery was an occupation which was more common at Bridgetown than in any other city in the British West Indies" (1980-30).

This subculture of prostitution was part and parcel of an informal domestic service sector that provided for not only the white planter class but also its maritime counterparts (Beckles 1989). Owners leased out slave women as a convenient way of earning money, and prostitution was considered more lucrative than the breeding of slaves. The sex market was very much linked to the more structured legitimate formal market, with the hiring out of female labor providing dual functions to individuals.

According to Beckles (1989) the more developed institutional patterns of prostitution were located in the Bridgetown area and linked to the leisure and entertainment facilities such as taverns, bars, and inns. It is within this context that the construction of different categories of womanhood along racial lines can be viewed where "white elite males possessed a sexual typology in which white women were valued for domestic formality and respectability, 'coloured women' for exciting socio-sexual companionship, and black women for irresponsible, covert sexual adventurism" (Beckles 1989:146).

Prostitution, in its traditional role, provided a means of support for not only black or colored women but also the white owners.

After emancipation in 1834 the practice was continued, indicating its entrenchment within Barbadian culture, with the role of prostitute/ mistress providing a steady income for those who worked either in a brothel or from home. Black men in this arena are given little attention. However, evidence suggests that their role within this subculture was very marginal, significant only in terms of offering "stud" services to the black females on the plantation, formally or otherwise. Hence, coming out of postemancipation society a typology of sexuality developed based on race and ethnicity, where the black woman was viewed as whore, the white woman as the Madonna, and the colored woman falling somewhere in between. The black man's role continued to be defined as marginal, except in terms of his sexuality, which came to be viewed as a defining characteristic.

Currently, although prostitution is an illegal activity, female prostitution is tolerated within the confines of established districts of Bridgetown, such as the historical red-light area of Nelson Street. It is deemed a female activity catering more to local demand, with the majority of prostitutes being streetwalkers and club workers. Little, if any, attention has been placed on male prostitutes as playing a significant role in prostitution in Barbados. The focus of this research is to fill part of this void.

MALE TOURIST-ORIENTED PROSTITUTION IN THE CARIBBEAN

Contemporary studies on tourist-oriented prostitution in the Caribbean have maintained that the phenomenon is based on racial stereotypes and sexual fantasy. Theorists focusing on Barbados have argued that "beach hustling" is based on the acting out of sexual racial fantasies about the natives (Press 1978, Karch and Dann 1981). In the case of male prostitution, an adherence to this approach would maintain that "hustling" should be viewed within a framework of continued First World exploitation of the Third World. However, to some extent, Press moves away from the view held by anthropologists and adopts a more plural framework, arguing that "although Barbadians tend to stress die sexual nature of tourist-hustler relationships,... the hustlers usually emphasize the material gain" (1978:114).

Considering sex tourism in Jamaica, Pruitt and LaFont (1995) contend that gender becomes a significant variable in analyzing these relationships between "Rent-a-Dreads" and the female tourists. There is "playing off of traditional gender repertoires" between the actors with regard to masculinity and femininity. They offer the term "romance tourism" to distinguish their views from those of anthropologists who deemed these relations as exploitative. They define romance tourism as a relationship "where the actors place emphasis on courtship rather than the exchange for money and sex. These liaisons are constructed through a discourse of romance and long-term relationship, an emotional involvement usually not present in sex tourism" (1995:423).[2]

O'Connell Davidson's 1996 study on sex tourism in Cuba, although presenting the obverse with regard to gendered actors, provides us with an insightful analysis about motivation and typologies of tourists involved in the phenomenon. Although this study adopts the more traditional approach of early feminist writings, the conceptualizations cannot be ignored, especially considering sex tourists.

The above studies provide us with a conceptual and theoretical base in which any work on tourist-oriented prostitution or sex tourism can be carried out. This study attempts a more eclectic theoretical framework in viewing male tourist-oriented prostitution in Barbados. Its special emphasis is on the definitions of the phenomenon by both actors, utilizing both traditional and contemporary approaches.

BEACH HUSTLERS IN BARBADOS

Beach hustling falls under the broader category of hustling as defined by pryce (1979) and Brathwaite (1983) as an alternate means of earning a living. In their sexual behavior, beach hustlers can be defined in terms of "gigolos'—men who receive material compensation for the social or sexual services they render to women" (Press 1978). In the case of Barbados, hustling can be viewed as the exploitation of white female tourists by young black Barbadian males. Beach hustlers are young, underemployed black men who provide sexual services and act as escorts to white female tourists of varying age in exchange for economic goods and services, which range from brand name clothes to airline tickets. Although these liaisons might start off as "sex for money" in the one extreme, if continued, emotional attachments are formed and the relationship is extended over a period of time, sometimes resulting in marriage.

The beach hustler, beach boy, or beach bum, as he is sometimes referred to in Barbados, is usually between eighteen and thirty-five years old and from a low socioeconomic background. He is not educated beyond the secondary level and has few qualifications, hence his subsequent inability to procure much in the way of jobs, except unskilled or semiskilled manual work, i.e., odd-jobbing, fishing, or painting. He might have started off as an odd-jobber working on the beach and procured more formal employment by assisting with the handling of jet skis or catamarans, or have a more entrepreneurial activity such as selling coconuts, coral, or aloes. As a respondent states, "I got into it as soon as I left school, 'cause I couldn't find no job, so I do something that could support me, but as time goes on you pick up different things. You pick up different ideas on how to do things...." Or as a key informant describes another beach hustler, Terry know a lot of white people. Terry does paint and do different things, while doing these things you meet people...." Hustling is not his only means of support, as a key informant articulates: "The hustlers are those who work with the water/ ski/ wind surfing and water sports. Most hustlers does work 'cause if you ain't got no money you does got to get yourself an income to buy yourself a drink."

This demographic profile of Barbadian beach boys parallels that given by Pruitt and LaFont of Jamaican Rent-a-Dreads, whom they describe as "a group of rural young people with little education and few social and economic prospects" (1995:428). Take, for example, twenty-one-year-old Carl,[3] who attended one of the newer secondary schools and left with no certificates. He usually supports himself by doing odd jobs such as carpentry and just used to "lime"[4] with the fellas on the beach, where he met Beth, a French European woman. She supports him when she is in Barbados, and he survives on remittances and odd fobs when she returns to her country of origin. Physically, he is well built, with well-defined muscles. He is very dark, obviously from his days in the sun, and sports short "natties" (short dread locks) which are bleached by the sun—a hairstyle that he is sure to tell you is "easy to manage, and the tourist women like it so."

Pruitt and LaFont make the same assessment with regard to hairstyle where "those men with dreadlocks who are assumed to be Rastafari receive substantially more attention from foreign women than do other Jamaican men" (1995:430). Although the Barbadian hustlers do not adhere to the Rastafarian characteristics, as their counterparts do in Jamaica, they do recognize the appeal of the hairstyle. As a beach hustler explains:

> You know why some of the girls like the knot-up hair? When some girls send photos and stuff up to England, you don't be seeing clean-cut men. They send a picture of Rastas. So when a girl come down here they think a Rasta is a real Caribbean man, so that is why they go for a Rasta man. Some girls like nature, so they say a Rasta like peas and vegetables. But some of them does fool themselves, 'cause they dont get a Rasta, they get an impostor. That's how they judge the Caribbean by the Rasta, and the thing is now when they come they can't go to Harbor Lights,[5] a Rasta can't go to Harbor Lights...."

However, the point must be made that in Barbados, the natties have become a trademark of the marginalized youth, which, it can be argued, parallel the Rastafarian appeal of the 1960s, so the wearing of the natties is a combination of a marketing strategy and a show of brotherhood.

The beach boy's attire is usually brand-name beach shorts worn in a manner to show off his well-endowed proportions. His appearance, an obvious marketing strategy, is based on the Western female's notion of the quintessential hypersexual black male—skin darkened almost blue-black to acclaim his pride in his African ancestry and to suggest an untamed, primitive nature and an exotic appeal. To the Western female he becomes the archetype of black masculinity (Pruitt and LaFont 1995).

The beach hustler tends to emphasize his masculinity, a point that reinforces the racial stereotypes of blacks, one of the exotic Other, in order to give the female tourist what she is expecting. The exotic Other has been constructed as more passionate, more emotional, more natural, and more tempting than his white counterpart (Pruitt and LaFont 1995). For example, when he goes to a club, immediately he will begin showing his "natural rhythm," gyrating to me latest calypso and reggae tunes, sometimes showing his "natural" athletic ability with a few seemingly improbable flips and splits. When the music changes to a slower beat, he will hold "his woman" and sing in her ear showing his "natural" ability to sing and demonstrating that he is "passionate" about her. Unlike his counterpart in Jamaica, he does not construct a "staged authenticity" (MacCannell 1989) but these are simply traits of his constructed masculinity of reputation (Wilson 1969). He is the black man that the white female imagines in this constructed paradise. The beach boy also has a large repertoire, including the ability to speak a sprinkling of a few European languages. Being able to communicate with his woman effectively is important in order to impress upon her the fact that he cares enough to try to learn her language. Pruitt and LaFont also note that in Jamaica, "learning to speak a little German or developing an expertise for guessing what types of experiences the specific tourists are seeking" also occurred (1995:431). Not only is the hustler skilled in languages, but his ability to speak in the accent of any tourist that he encounters is exceptional. Further he has quite a cosmopolitan lifestyle, spending a few weeks a year in Germany or Sweden with "a friend."

The young male is quick to deny that his liaisons with white female tourists are hustling or prostitution, as it is known in the Barbadian context. He points to me fact that "hustlers don't work and workers don't hustle for white women." Rather he seeks to make the clear demarcation between his "situational liaisons" and what he calls the "gigolos," "them guys who go around fucking and harassing the tourists for money." The interviewer, however, could not identify this category in the course of the research. For the hustlers, situational liaisons might start off as any other between male and female: "Yah meet a girl on the beach or at a disco, and she start smiling at you, and you start smiling at she, and you'll get to dancing...." As Carl asserts:

> The beach ain't no prostitution thing like that, it's a work and during yah working you get involved with people from all over the world and some like you 'cause they friendly and to compare with the people that from your country they more nicer, they more friendlier, they even help you more than you own, so yah get kind a close to them and all kind a scenes play.

Yet at the same time, these situational encounters are selected on the basis of one primary prerequisite, i.e., wealth. Tim, for example, asserts: "I like to get involved with executive ladies, with women with class, women with cash. Technically speaking I love women who have money. I can tell the ones who got money...." His partner also iterates:

> As a matter of interests if you meet someone, and they say they like you they will return shortly, like next week. That's how you find out if they have a lot of zeros behind the point. Not many people can do that if you save for the vacation....

They even speak of sometimes being fooled by the female tourists, who pretend to have money in order to gain their attention.. "Some of them girls does come down here and don't got nothing but credit cards and fool the boys. You got more than them. You see a girl at a bar with a credit card down and you think that she got money...."

Economics more than romance seems to be the mitigating factor in these situational relations on the part of the beach boy. They too, however, refer to the beach boy or beach bum at times unthinkingly to denote the relationship between young black males and female tourists, like themselves. For example:

> All the guys working in the bank hustling tourist women too because they see the fellas getting ahead, progressing quick, so they let me get in on this action. When I go to the Boat Yard[5] checking the women, a lot of businessmen in there. Is not only on the beach, you know....

Tim admits that to some extent that he defines himself as such

> You know that I is a beach bum, because when I go down to Speightstown every day of the week I walk with two or three grand in my pocket, and a lot people can't get that and I working legal. So, like to *be* a beach bum. I ain't inferior, I ain't dirty. They got land bums too you know; a lot of girls is land bums, go bum a man. See men with heavy with big gold, and see him and say boy I have to get to know him....

THE WHITE FEMALE TOURIST

The literature on gender and tourism articulates that "travel has become part of the gendering activity of women as they seek to expand their gender repertoires to incorporate practices traditionally reserved for men and thereby integrate the conventionally masculine with the feminine" (Pruitt and LaFont 1995:425). Travel for the emancipated western woman becomes an arena to test out new notions of a liberated femminity that goes in quest of the sexual Other, an Other who is endowed with a primitive masculinity that can no longer be found in the West.

The female tourist in this study is between twenty-five and fifty years old. She is usually from Europe, maybe Belgium, France or England, but may also come from the United States or Canada (see also CTO 1996, Press 1978, Karch and Dann 1981). She would probably have saved for this trip to Barbados. She is a secretary or nurse or holds some other job that puts her in a middle to lower income. She would tell you that she heard about Barbados through friends or acquaintances, generally female. Her first trip to the island would be with another female friend. Typologies of the white female tourists can be applied with regard to their liaisons with the beach boys. These are: the situationer, the repeat situationer, and the one nighter.

The Situationer

In this category, the tourist does not apply the label of hustling or prostitution to her encounters with beach boys. She emphasizes the romance element of the relationship. These relationships are situational encounters paralleling the typology of the situational sexual tourist of O'Connell Davidson's studies in Cuba and Thailand (1996). In this context she is defined as a female tourist who travels not with the intention of having sex with the locals but who enters into these relationships as soon as the opportunity presents itself.

An example involves Beth, who came with some friends, met George on the beach, and spoke to him a few times. They encountered each other in one of the many tourist spots, started dancing, and things developed from there. Beth bears the economic burden of the relationship, since George does not have a full-time job. When asked about his job status, the researcher was told: "I am not working right now, but I do this and that... Right now I just showing my lady here a good time." She, like her male counterpart in Cuba, denies the instrumental nature of her encounters, being quick to tell you that she is "not like those other girls who just come down here to have sex with a black man." But Beth like so many of the situationers tends to support her guy, and he stays with her in a guest house for the duration of her stay. The situation still remains very much sex in exchange for money and services, a situation that Beth seeks to ignore.

The Repeat Situationer

Another variation is the female tourist in her early to late forties who returns about two or three times a year. She is a well-known figure on a particular beach, which is near her guest house or hotel. She would have had a number of "boyfriends" over the years and have alternated between having them visit her country and traveling herself to Barbados. She tells you that eventually she wants to settle here as soon as she has enough money, because "everything here is so nice, and the people are so friendly." She too denies the remunerative nature of her relationships.

A case in point is Paula, who has been coming to Barbados since 1990 and has had three boyfriends, one of whom she has a child with. Paula speaks to me as she clings possessively onto the arm of her boyfriend. Responding to a query on why Barbados, she said, This is a very beautiful place and I found what I was looking for. He is everything to me." Her current boyfriend is around twenty years old, and he too has no full-time employment status; rather, he helps out the guys on the beach. Paula "helps him out" by buying him clothes and anything he wants, as well as sending him money to tide him over until she returns.

The One Nighters

Although the researchers were unable to access this group, its existence was quite apparent in the details provided by male and female key informers. These girls are usually British,"come down here to say they fuck a black man." The one nighters are usually in their twenties who are "here for fun. You will see them with one guy tonight and tomorrow night you see them with another. The next day you see them with another hustler...." These women behave similarly to the macho lads of Cuba, seemingly content with anonymous sexual encounters (O'Connell Davidson 1996).

THE DILEMMA OF ROMANCE, SEX, AND MONEY

The usual setting for encounters between tourist and hustler is the beach. This provides the backdrop for chance meetings, overtures, and sexual encounters and is the locale for this informal sector. As maintained by one key informant, The beach is a place where you have all kinds of hustling..." Clubs catering toward entertaining tourists also provide a backdrop for these encounters. These are the two settings where the beach boy can affect his "cool pose" and his "hustle."[7] The successful encounters are usually consummated on the sand.

The encounters are usually based on a mixture of racialized sexual fantasy, economics and emotion. Although, the researchers contend that the liaisons might start off as a combination of sex and economic services, they can be easily fitted under the umbrella of prostitution. The relationships may be extended, resulting in repeat visits to the island, visits to the tourist's place of origin, or by the migration of one of the actors involved. However, the motivation on the part of the beach boy usually is of the former variety, and their involvement in a steady romantic relationship becomes an indirect result of prolonged encounters, while the white female tourists define these liaisons in terms of situational romances and racialized sexual fantasies.

A conceptualization can be made, with reference to the hustler and the white female tourists that is a mixture of motivation and circumstances:

1. Racialized fantasies—racial stereotypes of sexuality between the actors.
2. Arrangement—economic exchange with some degree of staged emotional involvement on the part of the male.
3. Mixed—both economic exchange and some degree of emotional attachment.
4. Emotional—emotional involvement or "love".

The researcher wishes to articulate that these categories are not mutually exclusive, but merely an attempt to conceptualise the relationship between the actors.

Racialized Sexual Fantasies: The Black Stud and the Easy White Woman

Liaisons between the hustler and the female tourist are based on mutual racial stereotypes of sexual fantasy. These are usually of the one night stand variety or, if extended, last no more than two or three days. These encounters are based on such stereotypes as "black men can go all night" and "white women are easy, and give you head, no problem." Both actors mutually realize some racial stereotypes. Take the case of Jake, who maintains that "most women if you meet them at the club, you can get a one night stand."

George also makes his position abundantly clear:

Bajan women can't fuck, and they doan even wanta suck you. You got to beg she to do it, and still she might not do it, and if she do it she acting like if she doin you a favor. Now a white woman, you gotta beg she to stop.

Arthur makes a similar claim.

Bajan women is too much trouble, you gotta feed she carr' she out, and buy things for she to eat and drink, but a white girl you just gotta tell she ooh you look brown today, and she would gi' you it just like that.0

For others

Bajan women got too much pride, you would find that a Bajan woman come to the beach to swim, they have nice bodies, they have a nice bikini on, and still they go in the sea with a long dress, why is that? What are you hiding? You don't want to attract the boys, that means that you either gay, or you like girls. But you don't want the boys to see your body. You buy expensive swimwear you go to the beach right, you look good in it, but still you got a tee shirt covering it up-But you find a tourist woman as old as eighty years old, looking terrible with a big belly, no ass and still showing her body.... I think that fellas just getting fed up, trying to ask a woman out, and just walking with her face in the air. Tourist women are not like that....

Female tourists also adopt similar attitudes. For example, Judy maintains.

It's an entirely different situation. Black men like fucking, black men enjoy the sex act, they don't make love. Some white men make love. Black men in my experience don't make love, they fuck, and it's all of this brag stuff. How good the act was, and how good they are, and it's all of this brag stuff....

One beach boy, in explaining the attraction by white female tourists to beach boys, speaks about the white male's inability to sexually satisfy his female counterpart:

They ain't up to the mark, so when them women come to Barbados and see that we black boys healthy and took good and thing. They [white women] want to try something new. You can't beat that I have seen women divorce their husbands, get divorce, you know what I mean....

Another speaks about the fontasy element

White woman come from overseas come to live out a fantasy and walk away; a man might take advantage of that situation.... Some women maybe telling themselves you see me for right now I like these jungle pictures, all of these big, strong men. I want to go down there and get fuck in the bush and thing, all kinda fantasies....

Actors adopt the racial stereotype of the Other. The western female embodies the free sexuality" and adventurism of the North in contrast to the "respectability" of the Caribbean woman (Wilson 1969), where open sexuality is frowned upon. The Barbadian male embodies the primitive, aggressive nature of the black man, having an animal-like quality, which is constructed as opposite to the white man, the gentleman. This is the type of man who "would fuck you in the sand and wouldn't think anything the matter with it"

The Arrangement

This type of relationship usually lasts the duration of the tourist's visit. The woman is approached on the beach while sunbathing or at a nightclub, and conversation is initiated about her visit, her impressions of Barbados. The hustler might then give himself the role of guide and promise to show her "the true Barbados." The woman usually becomes enamored by such attention, which is further reinforced by his propensity for "sweet talk" (Abrahams 1983, Wilson 1969) and agrees to the hidden layers of the "contract". The arrangement is usually sex and money or goods in exchange for such services as tour guides and escort to clubs and restaurants.

In the arrangement, the hustlers emphasize the economic aspect of the relationship. Peter points out:

> Tourists carry you to dinners. when I was on the sand [beach] I used to eat food at Sandy Lane,[8] and all today, I still eating food there me and my girl at Sandy Lane. I don't spend anything. I don't get to spend a cent. They invite me out, they going to treat you nice....

Ted makes the claim that "Bajan women always looking for somebody to support them. I looking for somebody to support me too, what is wrong with that? I happy." One key informant puts it more dearly: "I like to get involved with executive ladies.... My women are lawyers, own their own companies, are executives of other companies.... Technically speaking I love ladies who love money." In fact, most of those who were interviewed had their own cars, were renting apartments, and owned or had part ownership in beach sporting equipment. All had been funded or given start-up capital by their girlfriends.

Female tourists within this arrangement classify themselves as victims of love, seemingly unable to see this relationship as an arrangement instead of a romance. However, at the end of the vacation, they recognize the reality. Susan, now living in Barbados, examines the plight of the female tourists:

> They get here and all of a sudden they are on a beach wearing next to nothing, and up and down the beach parading are all of these gorgeous men like peacocks coining up to

you saying, "You are so beautiful, you are so gorgeous. Can I take you out tonight, can I do this, can I do that". Before you know it the girls are intoxicated.... The usual scenario is this: she spends her money on him, she pays for the taxi, she buys the clothes. She mistakes sex for love....

James explains why they are mistakenly caught up

They does get into us and do everything with us, every-thing. We does carry them places, island tours, sailing and so on. We does be with them showing them everything, giving them a good time, making sure they enjoy their holiday. So they are going to devote time with us, they ain't going to look for another person....

Pruitt and LaFont (1995) also found that in the Jamaican context where the woman is caught in a relationship, its nature was not initially defined in this context.

Mixed/Romance Relationships

Mixed relationships might start off as arrangements, owing to continued contact between the two actors. The nature of the relationship begins to develop along romantic lines, although the economic aspects still play a significant part. The hustler finds it more lucrative to stay with this particular woman, and on his part begins to develop feelings for her. This type of relationship closely parallels the romance tourism framework of Pruitt and La-Font (1995), where the courtship element in the relationship begins to play a more significant role. It is in this case that the hustling element of the relationship is usually threatened and the cool pose begins to crumble. Like Aldrick the Dragon Man in Lovelace's novel (Lewis 1998), it calls for a redefinition of masculine identity "to be more himself than the pose [or the hustle] could ever be" (1998:172). According to Mark:

At first, we weren't in anything, you know; she used to come down here and would show she a good time, then she start staying longer and longer.... Now only got she one, and I does go up and she does come down.

James also holds a similar view. He blames it, however, on sex: "People fall in love despite not wanting to. Sex is a powerful thing. You start off, you want a fucking, before you know it there is feeling there.... Maybe it's the pussy, maybe it's the lady...."

The women also develop deep feelings toward the young men. Beth, a Belgian who has had a relationship with Rick for two years, maintains that "after a while, I started to love him, I started to think about him constantly, and I would call him, and when I came back he also had feelings for me."

Emotional

Usually if the mixed relationships endure, they continue to the next level, where marriage can result. One of two things can happen. Either the hustler emigrates to the woman's country of origin or the woman moves to Barbados. In the North, the marriage often ends, since the constructed world cannot be maintained in the reality of living day to day in the female's country of origin (Wagner 1977). Hustlers usually complain about the weather or the fact that they felt constrained. "She didn't want me go nowhere, only with she...." Female tourists on their part are surprised when their new husbands continue along the lines of the stereotype of the black man, for example, indulging in promiscuous behavior, obviously not realizing that this is part of the beach boy's masculine identity. As Paula maintains: "He had a huge addiction to chasing pussy...."

Those marriages that survive do so when the female decides to live in Barbados. An example is the marriage of Meg and Dave. Meg works as a tour rep, while Dave works at a windsurfing club; they just had a baby. Both declare that they are quite happy together; Meg even speaks with a quasi-Barbadian accent. Meg confesses that she still has some problems with her adjustment to Dave's family and to Barbadian women in general, who are highly suspicious of her. But for the most part, the couple still has a lot of friends, mainly their co-workers. The relationships that develop between the hustler and the female tourist tend to foster along the periphery of the tourist culture. Seldom do they actually enter the local culture.

NEGOTIATING GENDER IDENTITIES

Barbadian working-class young men are products of a colonial legacy, which has fostered the marginality of the male, both socially and economically, in society. Caribbean writings have articulated the complexities of Caribbean peoples' inability to adopt Western European gender roles within the context of poverty and deprivation (Clarke 1957, Massiah 1986). The result has been the dominance of Caribbean women in the family and in the marketplace, while Caribbean working-class men have been denied equitable access to competition in the market place by virtue of institutional racism and have continued to be unemployed and underemployed in the face of economic recession.

Black masculinity is constructed as a response to the colonial framework of the white hegemonic masculinity. It is a form of resistance and an active redeployment of white masculinity (Cornwall and Lindisfarne 1994). It is measured in terms of male virility, which is manifested through sexual conquest and the fathering of many children—poor man's riches—and established by other behaviors, such as boasting in the public domain and exaggerated complementing or "sweet talk" (Wilson 1969.106).

Beach hustling in this context represents an attempt by young black males to construct a type of masculinity whose prosaic symbols are achievable within Caribbean society. Tourism has facilitated development of an informal sector in which "he is his own man," not working for anybody and answering "yes sir, no sir." In fact, this entrepreneurial activity allows him to affirm his masculine role among his peers. Beach hustling provides the forum for the young black male to successfully test the limits of a cultural masculinity. He uses cultural constructions to his economic advantage in his liaisons with female tourists.

The white female, with her foreign income power, adventurism, and race, assumes a hegemonic gender power in Barbadian society denied to the local Barbadian woman. Her encounters with the beach hustlers represent an opportunity to test out this hegemonic dominance under balmy skies in the center of the Other, with the Other, who, she maintains, is "so different from the men at home... so aggressive" (aggressive in his "come on" tactics and his general demeanor). She adopts the role of economic provider within his cultural context She understands that "her guy is from the ghetto.... He doesn't have any fucking money..." and that is all right by her, because he shows her a good time, becoming her passport to Barbadian culture. She feels guilty about her comparative economic wealth and overcompensates by not only paying for him to go places but also buy him clothes and anything that he may need: "She does what she can for this poor guy living in the ghetto."

Unlike his Jamaican counterpart, the Barbadian hustler has no problems dealing with the economic dominance of his white partner, since this role is not new to him in the context of the local culture. The status and recognition given to him by his peers subsume any contradictory feelings that he may have (Press 1978). His status is augmented by the material gains of the relationship, not only by owning jet skis or brand-name clothing but by the propensity of the woman to provide him with opportunities denied to him by his race. As one key informant maintains:

> A woman would send for me from here to go overseas with a return ticket. I can travel from here any part of the world. I can make a phone call and say I want to leave Barbados they would say, "Come, the ticket on the way." I don't have to ask twice you understand...

The white woman does not threaten his masculinity by demanding symbols not constructed as part of his cultural masculine role, e.g., money—demands that are so much part of his relationships with local women. In the context of a Caribbean culture, it is expected within a relationship that the man will give money to the woman (Pruitt 1992). As one informant asserts, "Bajan women want too much material.... Big man with big car.... White woman don't check for nothing so...."

Further, these encounters provide opportunities for the black male to affirm and reinforce his cultural masculine role of sexual prowess, and his other "expressive" traits like sweet talk. He is also able to demonstrate his "skills of strength and knowledge" (Wilson 1969:106) in his role of tour guide and escort. She also allows him to be "a man" and to adopt a dominant role in the relationship, which is usually denied in encounters with local women. According to George: "You gotta to be tough when you dealing with these women. Not fight with nobody or nothing, but you gotta know what's going on around you, if you versatile with the language you pull women easy...." Another key informant maintains, Them done know that we are kings. We are kings, creatures of the earth you understand De white boys are more laid back and take de orders, de black boys now does give them de orders...."

In a postcolonial society where race, color, class, and status are very much intertwined, and with society having a predilection to accord status based on color, being seen with a white automatically accords her black companion some status within the local society (Lowenthal 1967, O'Connell Davidson 1996). If the analysis of Karch and Dann (1981) is accepted, these relationships give the black male an opportunity to act out his own fantasy of being a white man and having the power that whiteness brings. It echoes Fanon's analysis:

> Out of the blackest part of my soul, across the zebra strip-
> ing of my mind, surges this desire to be suddenly white. I
> wish to be acknowledged not as black but as white.... who
> but a white woman can do this for me? By loving me she
> proves that I am worthy of white love. I am loved like a
> white man, I am a white man (Fanon 1970:63).

Arguably this can be conceived as being true in this context, since many of the beach boys interviewed no longer have relations with local black women. The interracial encounters provide a context for the liberation of masculinities denied to them in encounters with Barbadian women. However, the point is noted that Barbadian women also refuse to have relations with those men who are labeled as hustlers. The hustlers also speak of Bajan women as having too much "pride," too much "attitude." The impression given is that in their new role, Bajan women are viewed as lacking in some respects when compared to the white female tourists who are viewed as more sexy, more adventurous, friendlier, a view that is in keeping with this still very racist postcolonial society.

The hustler's companion, in her hegemonic feminine role and her quest of the exotic Other, is not unlike her male counterpart and is able to construct a gender identity denied to her in her own country. She adopts the masculine role of economic provider yet also explores the traditional feminine role thrown off by the feminist movement. It allows her to explore these gender contradictions, playing both man and woman, in the center of the Other. She seeks out the quintessential male with his attributes of primitive sexuality and sexual prowess that is denied in the construction of the "western male, and in her new hegemonic role she has the power to do so in paradise—the stuff of romantic novels. As one informant argues, "It's a combination. It's the island. It's being away from England. It's the air, the warm blanket around you, the feeling of the balmy air on your skin and them coming along telling you how gorgeous you are...." In essence, these encounters present an arena where the western female, although adopting some degree of the masculine role and hence having some measure of power, can enjoy the traditional role of femininity.

THE ROLE OF MIGRATION

Migration has been given very little focus in tourist-oriented prostitution in the Caribbean, with the exception of Press (1978) and Pruitt and LaFont (1994), where it was found that tourist-oriented prostitution allows the local "prostitute,'" by virtue of forming extended liaisons with the tourists, to emigrate to the "Center." Historically the Center has been perceived by the Other, in colonial and postcolonial ideology, to be "El Dorado"—a place of endless opportunities. These interracial encounters allow the beach hustler to fulfill this dream of endless opportunities in the context of limited available options within the host culture. Indeed, as articulated by Pruitt (1993:147), in the context of Jamaica, "hooking up" with a white woman becomes the passport to a better life. In Barbados a similar situation arises, where many of the beach hustlers speak of "the boys who get thru"—who have gone on to a better life—since "Barbados ain't got nothing to offer you." This becomes the ultimate goal for many, and those who are successful are admired by their peers.

A case in point is Trevor, who appears as the prime example to other beach hustlers. Trevor met a rich woman from Germany, whose family is in the construction business. Now he divides his time between Barbados and Germany and when he returns exhibits all the antics and symbols of a rich man, including a white Mercedes. Indeed, many of the guys interviewed spoke of spending a few weeks in the United States or Europe. They explain that "sometimes a guy might meet a woman who has got money, he can go up there and see what he can do to get an opportunity, because a guy wants a break."

CONCLUSION

Beach hustling in the Barbadian context is a social and economic activity that provides an opportunity for marginal young black men to "achieve some notoriety, if not self-esteem, in their home communities and male peer groups" (Press 1978:116). It provides the young black man with a forum in which his cultural definitions of masculinity are affirmed. It allows for successful negotiations of gender between himself and the female tourist. He is able to achieve status, material goods, and independence through a legitimate but wholly unconventional set of means in the context of the wider society. In fact he is quite adamant about the significant role he plays in the tourism industry and views beach hustling as "playing his part," sometimes more effectively than many others. The beach hustler embodies all the racial sexual stereotypes that the white emancipated woman has come in search of. Tourism allows the consummation, realization, and affirmation of these two gender identities within this paradise.

NOTES

1. The research was carried out over a period of six months in the latter part of 1997 and early parts of 1998 by the researcher and an assistant. The main techniques of investigation were observation, informal in-depth interviews, and focus groups. Observation was carried out at the main "pick-up sites," namely popular West Coast beaches and main tourist clubs. Respondents were selected on the basis of consent and were asked for the names of additional potential informants. Interviews were conducted with twenty beach boys and ten female tourists, the latter proving quite difficult to access. Three focus-group sessions were also conducted. Additional interviews were conducted with individuals who, though not engaged in tourist-oriented prostitution, could provide useful perspectives for understanding the phenomenon, e.g., hair braiders, security guards, and hotel workers.

    The researchers recognize the existence of other types of prostitution, of which hustling of female tourists by young black males is just one type. Observation suggests that female club prostitutes as well as streetwalkers cater to male tourists as well as the local clientele. Further, female beach hustling, very much like its male counterpart, is also noted as specifically catering to white male tourists. The researchers also observed male homosexual hustling of male tourists. However, the focus of this study was intended to be the liaisons between black male Barbadians and white female tourists, thus male-male relationships were not subject to further examination.

2. To the extent that there are "transformed gendered" identities and contradictions of "conventional notions of male hegemony" on the part of the Jamaican males speaks to the ethnocentric bias held by the researchers with regard to theorizing of masculinity. Their tendency is to assume that the 'western notion of hegemonic masculinity is the same in the context of working-class Jamaica. However, even with these limitations, the study does provide a useful frame of analysis from which to view other studies on male tourist-oriented prostitution in the region.

3. Any names mentioned in the text are pseudonyms owing to the guarantee of anonymity.

4. Barbadian term for a social gathering or for hanging out.

5. An exclusive nightclub in Barbados known somewhat for its racist policies.

6. A popular tourist club in Barbados.

7. Hustling or "cool pose" are responses to the young black man's alienation from the confining trappings of traditional masculinity (Pryce 1979, Majors 1986). The evidence suggests that hustling itself can connote any type of informal job activity, illegal or otherwise. Taken in this context, hustling and cool pose are cultural constructions of black masculinity, responses that concomitantly rise in direct relation to unemployment and economic hardship.

8. A very exclusive hotel on the west coast of the island.

CHAPTER TEN

Fabricating Identities: Survival and the Imagination in Jamaican Dancehall Culture

BIBI BAKARE-YUSUF

INTRODUCTION

An assessment of recent work on Jamaican dancehall culture reveals the absence of any systematic analysis of the role that fashion and adornment play in the culture. This is surprising given that fashion is a prominent and constitutive part of the culture *and* the site for vigorous debate about lower-class women's morality and sexuality in Jamaica. This failure can only be attributed to the fact that analyses of dancehall culture have generally focused on lyrical content, the sound system and the economic production of music (Stolzoff 2000; Cooper 1993a}. I suggest that underlying this focus is the implicit assumption that music equates with inferiority, language and "deep" meaning. In contrast, adornment and fashion are considered to elude or even destroy meaning. Therefore, to invest energy on adornment conjures up images of superficial, transient and frivolous activities undertaken only by women (Polhemus 1988; Tseëlon 1997), in contrast to the serious male world of ideas connoted by music production. While undeniably significant for a critical analysis of dancehall culture, a continued over-emphasis on music and lyrical content, to the neglect of other aspects of the culture, unwittingly privileges the activities of men and their interpretation of the culture to the exclusion of women.[1]

In this article, I want to shift attention away from lyrical content and a specific focus on Dancehall culture to examine the embodied practices that emerged in the late 1980s to the end of the 1990s in Jamaica. I argue that working-class Black women in Jamaica use fashion to fabricate a space for the presentation of self-identity and assertion of agency. Through adornment, dancehall women have been able to address creatively the anxiety, violence and joy of daily life. At the same time, they have been able to register historical, cultural, economic and technological changes through their bodies (Breward 1994). Prior to speech or any written manifesto, different modes of adornment are employed to contest society's representation of and expectation about lower-class leisure activity, morality and sexual expression. In a nutshell, fashion allows dancehall women[2] to challenge the patriarchal, class-based and (Christian and Rastafarian) puritanical logic operating in Jamaica. Of course, the wider context in which this articulation of social relations has taken place is that of socio-political and economic realities which includes continued anti-black racism, black nationalism and global cultural and economic restructuring. In this sense, far from fashion being meaningless, superficial and unworthy of cultural analysis, it allows working-class black women to invest their everyday lived realities with multiple meanings and processes which links them to both the spectacular fetishism of global consumerism and mass media semiosis as well as the African love of ceremonial pomp and pageantry.

The fact that dancehall women most often do not consciously adhere to this critical position in their speech, nor readily perceive their action as jamming the hegemonic syntax, is quite beside the point.[3] Phenomenology teaches us that there is often a gap between intentional action and explicit, self-aware interpretation (Tseëlon 1997). Far from imputing a kind of rational, contestive voluntarism to dancehall women, I suggest that the significance and meaning of their action as a form of contestation is not always available for self-articulation. As such, my account and interpretation of the meaning of fashion and adornment in the culture is not wholly circumscribed by empirical enquiry into conscious explanation, speech acts, or verbalized discourse. Rather, my analysis is based on *both* empirical engagement and my own analysis of the expressive body in the culture. This body, its desires and perceptions are seldom fully disclosed within speech; rather, they are made manifest in a variety of bodily practices.

CHANGING TIMES, CHANGING STYLES

Fashion styles are always embodied and situated phenomena, reflecting and embodying social and historical changes. Prosperity, crisis and social upheaval are stitched into the fabric of every epoch. For example, the extraordinary wealth of Renaissance Europe was materially layered into the ornate and elaborate detailing of upper-class clothing. In contrast, in its rejection of the sumptuous and colourful style of the *ancien régime,* post-revolution France adopted a less ostentatious and simple cut in order to reflect newfound freedom (Breward 1994; Connerton 1989}. Fashion and bodily practices became a crucial site to both express wealth in the one case and challenge old hierarchies in the other.

Among New World Africans, fashion and bodily practices have also absorbed and expressed key symbolic functions. Starting from the long revolution fomented in the hold of the slave ship, when the enslaved cut their hair into elaborate designs (Mintz and Price *1976)* and reaching its apogee in the Black Power movement of the 1970s, bodily practices were as important as political manifestos in the struggle for freedom, agency and assertion of cultural identity. The "natural" Afro hairstyle, dashiki, large-hooped earrings, psychedelic skirts and patchwork miniskirts of the 1960s and 1970s signalled a rejection of European aesthetics. Fashion styles visually represented and extended the ideological affirmation and valorization of blackness and Africanity in circulation during the period. Like their North American counterparts, many Jamaicans adopted Africanized textiles, kaftans and long flowing brightly coloured majestic robes, head-wraps and jewellery made from natural materials such as seashells. However, during this period, the most important challenge to the aesthetic and ethical sensibilities of the Jamaican elite came in the form of Reggae music and the wearing of dreadlocks. Many Rastafarians and Reggae fans adopted dreadlocks and the military uniform of khakis and combat trousers. These motifs not only posed a challenge to the white capitalist and Christian ideology pervasive on the island, but they also drew attention to the permanent state of warfare that characterized life in the downtown ghettos of Kingston. Rastafarian fashion, in particular the wearing of dreadlocks, performed a critique of the dominant regime, asserting an alternative cultural, ethical and aesthetic sensibility in its stead.

230 | CHAPTER TEN

In the late 1970s, the rise of Edward Seaga's neo-liberal free-enterprise government heralded a new era of increased insecurity, violence and anxiety. This political turn gave rise to a corresponding cultural energy. Popular cultural expression on the island such as music, dance and clothing style moved away from the socialist, pan-Africanist eschatological project associated with the Rastafarian Reggae of Bob Marley and Michael Manley's socialist government towards what appeared to be the hedonistic, self-seeking pleasure and excess of dancehall style (Chude-Sokei 1997, Barrow and Dalton 1997). Because the elite found it offensive and it was not initially given airplay on national radio, dancehall music was generally produced and consumed in the open air or indoor spaces designated as dancehalls. Dancehall music or ragga (as it is known in Britain) is reggae's grittier and tougher offspring, making use of digital recording, remixing and sampling while DJs "skank" or "toast" over dub plates (Jahn and Weber 1992). To the ruling elite, the music was considered pure noise, a cacophonous drone that grated the nerves. The lyrics were considered bawdy, guttural and sexually explicit. Finally, the elite considered dancehall (especially female) dress and adornment brash and excessive, reinforcing the view that lower-class Black women were sexually permissive. As a whole, the subculture confirmed to the Eurocentrically inclined elite the immorality and degeneracy of the urban poor.[1]

In contrast to this disparagingly reductive view, dancehall culture should be viewed as a complex reminder of the continued relationship between popular expressions, commodification, urbanization, global economic and political realities and historico-cultural memory. For example, in music, older Jamaican styles and practices were revived and brought into conversation with new digital technologies and global flows of information (Bilby 1997). In terms of dance, dancehall unearthed older Jamaican forms such as Dinki Mini and Mento. As the Jamaican choreographer and cultural theorist Rex Nettleford notes,

> The movements in dancehall are nothing new; in my own youth I witnessed and participated in men to sessions which forced from executants the kind of axial movements which concentrated on the pelvic region with feet firmly grounded on one spot (1994: 1C).

Dancehall fashion fits into a general "African love of pageantry, adornments and social events...". (Mustafa 1998). The African-American folklorist Zora Neale Hurston suggested that "The will to adorn" constitutes "the second... most notable characteristic in Negro expression" (1933: 294). The will to adorn, she argued, is not an attempt to meet conventional standards of beauty, but to satisfy the soul of its ' creator (ibid.: 294). I suggest that the desire to "satisfy the soul" and project their own aesthetics onto the world is at the core of dancehall women's sartorial practice.

DANCEHALL FASHION

In a society influenced by Christian Puritanism and the sexual conservatism of Rastafarian ideology, dancehall fashion has responded antithetically with bare-as-you-dare fashion. Unlike previous African diasporic youth subcultures, dancehall is unique in that women are highly visible. Although there are a number of prominent female dancehall music performers, women's visibility in the culture centers on their ostentatious, sartorial pageantry. The "session," "bashment" or "dance" is an occasion for visual overload, maximalism and the liminal expression of female agency. Women form "modelling posses" or rival groups, where they compete with each other at a dance event for the most risqué and outlandish clothes. Their consumption practices are largely funded through the informal sector, as hagglers {informal commercial traders}, petty traders, cleaners, dancehall fashion designers or by having a "sugar daddy." Many of the outfits are designed and made by the wearer or by a local tailor. The style appears anarchic, confrontational and openly sexual. Slashed clothing, the so-called "lingerie look" (such as g-string panties, bra tops), "puny printers" (showing the outline of the genitals), Wild West and dominatrix themes, pant suits, figure-hugging short dresses and micro hot pants infamously known as "batty-riders" are favored. Revealing mesh tops, cheap lace, jeans designed as though bullets have ripped into the fabric and sequined bra tops became an essential part of dancehall women's wardrobe in the 1990s. At the close of the 1980s, the dancehall female body was wrapped in bondage straps and broad long fringes or panels attached to long dresses to accentuate the fluidity of the body's movement in dance. Incompatible materials and designs were juxtaposed - velvet, lace, leather, suede, different shades of denim, rubber and PVC, as well as animal prints such as mock snake, zebra and leopard skin, to produce an eclectic personal statement. Seemingly irreconcilable colours are combined to produce a refreshingly audacious, motile canvas on the dance floor. According to Carolyn Cooper, the sessions are the "social space in which the smell of female power is exuded in the extravagant display of flashy jewellery, expensive clothes [and] elaborate hairstyles" (Cooper 1993a: 155).

Hairstyle, make-up and jewelry are a key part of the dancehall look. In the late 1980s to late nineties, huge cheap and chunky gold earrings with razor-blade designs, as well as necklaces with dollar signs were worn on the ears, nose, nails, waist, and belly button as status symbols. More recently, the style has moved towards "ice" (slang for diamonds) and "bling-bling" (code for expensive jewelry and accessories). Hair is either dyed in bright colors or covered in metallic-colored wigs, weaves and extensions (platinum blonde, orange, turquoise, aubergine, pink). This style disrupts the Jamaican elite notions of "good" and "natural" hair versus "bad" and "processed" hair. In so doing, dancehall women draw attention to the artifice of African hairstyle (Mercer 1987) and the way "black women exercise power and choice" (Banks 2000: 69).

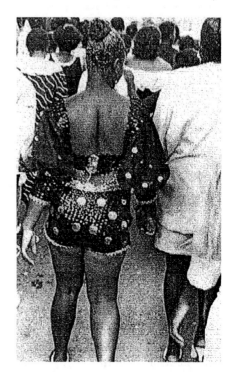

In opposition to the Jamaican elite preference for understated beauty characterized by lightly applied make-up highlighting flawless skin, dancehall women's make-up is deliberately bright, glittery and brash.

Shoe styles continue the sexual fetish theme of the clothes: laced or zipped up stilettos, knee or thigh-length boots in patent leather or "pleather" are favored for their emphasizing effects on the crotch and thighs. High-heeled strap shoes that coil round the calves towards the knees complete the image (D'Elia 2002).

DISRUPTING BEAUTY, CLASS AND GENDER

Given the intensity of dancehall modes of adornment, the question that arises is how to account for the phenomenon, both in terms of the socio-historical context of Jamaican culture and dancehall fashion's position within a global economy of signs and material flows. Here, I will identify three interwoven processes at work. First, I will point to the ways in which the fashion disrupts Jamaica's elite ideas of black female beauty and norms of appearance, via an overtly sexualized paradigm shift towards what I term the "aesthetics of voluptuousness." Secondly, I will show how practices of adornment in Jamaican dancehall are women's response to a "limit situation" in which the issues that face them on a daily basis are issues of survival and keeping the terror of daily life at bay. Contrary to the conventional (Western) impulse to assume that limit situations necessitate a victim mentality, attenuating agency and expression in the process, I will argue that limit situations can stimulate the collective imagination to heightened states of intensity and excess. Finally, I wilt show how dancehall fashion is located within a network of global flows and exchanges. Rather than being a uniquely local phenomenon explicable solely in local terms, dancehall fashion demands a broader understanding of how both local and international signifying systems have been adopted and appropriated by Jamaican women, creating the distinctive aesthetic form and self that is dancehall style.

In order to understand the ways in which modes of adornment in dancehall work to disrupt existing conventions of the beautiful and express sexual difference, we need to appreciate the Jamaican upper-class perspective against which it responds. In Jamaica, as elsewhere in the New World, the voluptuous black female body came to embody upper-class anxiety over the moral status of the lower class. According to Carolyn Cooper, it is this body that became the site for "ongoing struggle over high culture and low, respectability and riot, propriety and vulgarity" (Cooper 2000: 350). Just as in late eighteenth- and nineteenth-century Europe, where a distinction was made between the modest and chaste (typically upper-class body) woman and the vulgar and obscene (working-class and African body), a similar bifurcation takes place in the Caribbean. The European upper-class moral ideal also operates on the island with an accompanying aesthetic transfer. The ideal look for elite Jamaican women is essentially a *white* look: slender, restrained, calm, long flowing (straightened) hair and light or brown skin. The continued value placed on skin color as an important aspect of social mobility still exerts its own peculiar influence and particular form of violence on the psyche of Jamaicans (Douglas 1992; Barnes 1999). Although class is a critically important social vector in Jamaica, it is often articulated "through the idiom of colour shade and can reflect biologized notions of race" (Austin-Broos 1994: 218). Class, color/race and gender in combination constitute what Austin-Broos has called "Jamaica's discourse of heritable identity" (ibid.: 218). This discourse perpetuates and reproduces the cultural logic of plantation society, where upper-class European values and morality have dominated the socio-cultural and economic landscape till the present day. It is the worldview of the brown or non-black elite which features most prominently in public arenas such as newspapers, beauty contests and television, and in the political and economic life of a country where 90 percent of the population is black. Lower-class black women are accordingly derided as vulgar, uncontrollable and dangerous. To the ruling elite, the protruding belly, large dimpled buttocks and thighs squeezed into revealing batty riders marks dancehall women as indecent, morally repugnant and unproductive elements within society. Writing in the *The Gleaner*, Andre Fanon echoes the views of the ruling class:

> Dance-hall becomes a danger when the dance-hall syndrome is made into a way of life ... Dance hall ... under-develops our women who feel that they must learn to "wine" and "cock out" their posteriors as champion bubblers. Dance hall as a way of life emphasises the unproductive elements in society. If not channelled, dance hall will create a class of people which is incapable of doing anything productive. (1988:14)

In their purposely garish coiffure, slashed latex body suits and flashy gold jewelry, the women assert their distance to and non-conformity with the sobriety of Fanon's official culture and all that it represents: formal work and chaste morality. Importantly, these sumptuary practices mark dancehall women out as being unaffected by the lack and poverty that is a characteristic of ghetto life. According to Nettleford, dancehall fashion "all together spells for many a form of personal liberation from the strictures of a humourless existence which the hardships of poverty like the cloying satisfaction of affluence seem to impose on human beings at the opposite ends of the social scale" (1994:16D). Dancehall women contest the association of material poverty with the inability to care and attend to public self-presentation. As one participant in the culture succinctly puts it, "I may be poor and come from the ghetto, it don't mean I have to look shabby. A woman always has to look good even if it means spending her last dollar. Going to a dance is the time to dress up and let the world really see you as you are." As Paulette McDonald notes of another dancehall fan, "dressing up in expensive clothes sends a clear message to those who think that ghetto people are the scum of the earth: they can indeed set the fashion pace" (1993: 10). Dancehall attire suggests that without engaging in the arduous work of the formal economy, women in the culture still have at their disposal the money and creative resources that enables them to invest and participate in a fashion-based signifying economy. Rather than directing their labor towards the formal, bureaucratic structure, dancehall women work on their body as a 'canvas of representation' (Hall 1992: 27) so that the world can see them as they are.

Moreover, against the upper-class slur of idle unproductivity as described by Fanon above, dancehall modes of adornment should be viewed as a form of work in its own right — the work of creative resistance as a product of a playful imaginative and historical retrieval. In the context of a conservative and moralistic society, dancehall adornment invokes a visual subversion by returning to the subterranean sources of Jamaican folk culture, demonstrating a capacity to admit variation as a form of continuation. Of course, as I suggested above, this subversive historical recall does not necessarily conform to an intentional politics of conscious resistance. Dancehall female fans often have an unconscious pre-theoretical response relation to the ideology surrounding them, the variety of ways it is working through them and the responses engendered by it. As the French sociologist Pierre Bourdieu points out, "It is because subjects do not, strictly speaking, know what they are doing that what they do has more meaning than they know" (Bourdieu 1977: 79). This unconscious collective agency of bodily action resists theorization in the language of conscious articulation. As elsewhere, the female body in dancehall perceives and responds in its own way to forms of affirmation and negation in the world. Through this unconscious collective response, dancehall women contest Euro-hegemony as it is manifested in Jamaica. Whether or not these resistances and refusals could ever be so well articulated in speech or deemed important is quite another matter.

Indeed, beyond being a form of creative labor, I suggest that the work of the imagination has as its goal the disruptive parody of the Jamaican class structure itself. From a perspective influenced by Marx's theory of commodity fetishism, women's stylistic excess can be decoded as class struggle by other means. The function of creative strategies of adornment in dancehall is to destabilize the field of class distinctions in Jamaica. It is helpful, in this context to refer to the work by Jonathan Friedman (1994). Friedman adopts and critiques Pierre Bourdieu's analysis of distinction-making in order to show how the Congolese Sapeurs appropriate fashion styles (especially that of European haute couture) in order to disrupt the field of taste distinctions at work in Congolese society. For Friedman, it is precisely through appropriation, repetition and strategies of exaggeration and excess that the Sapeurs destabilize normative conventions of taste. By borrowing modes of dress and aesthetics associated with the economic elite, the Sapeurs (who are mostly poor and urban dwellers) undermine the commodity fetishism and prestige value with which exclusive European clothing was previously invested. His argument therefore borrows from and exceeds that of Bourdieu. For the French sociologist, taste distinction occurs within a field—a horizon of differences that comprise the differential fabric of society. For Friedman, in contrast, the work of adornment among the Sapeurs tears at the social fabric of taste. In a parallel manner, in dancehall culture, a far more destabilizing dynamic is unleashed than that of simple social differentiation. Dancehall women are less concerned about being accepted by or assimilated within mainstream society, or to pretend to be something they are not. Again, they are not concerned to mark out a simple difference in taste. Rather, the overloading of the senses through sartorial extravagance works to deconstruct and jar the field of distinctions itself, through the exaggerated mimicry of conventional conceptions of feminine comportment. In a forceful response to denigration from upper/middle-class sensibilities and the Rastafarian restricted "mother earth" style for women, dancehall women dress to oppose every aspect of "appropriate" feminine comportment, appearance and conduct. Whereas middle-class Jamaican women tend to desire a slender and sleek figure, dancehall women rejoice in an unruly voluptuosity—the joy of being fat. And more troubling, slender women have resorted to using drugs, food and even hormones, the "fowl pill," to get fat as described in one of Lexxus' lyrics.[5] In opposition to the Rastafarian chaste concealment, dancehall women revel in exposed flesh. Where both uptown and Rastafarian women value stylistic restraint (in terms of fabric, fit and colors), dancehall women value riotous colors and sheer maximalism. In this sense, the fetishized commodity of cloth in Jamaica, divided between elite, Rastafarian and poor urban-underclass sensibilities, threatens to break out of a rigidly Marxist perspective of class struggle to include a contestation of what it is to be a woman. Dancehall fashion is considered such a threat by the ruling elite that at the Emergency Department of the Bustamante children's hospital in Kingston a "Dress Code" is deemed necessary to restrict dancehall aesthetic from spilling over into such an institution:

I.   Please cover all body parts!!

II. Please attend to personal hygiene!!

III. No setters in hair!

IV. NO Dancehall Style! (cited in Cooper 1993b)

Dancehall culture therefore is the site of a contestation of both class and sexual difference, to the extent that neither can be reduced wholly into the terms of the other. While the aim of dancehall style is meant to shock and rebel against the upper- and middle-class and Rastafarian ethos, it cannot be fully understood outside of an attempt to intervene against repressive attitudes towards female sexuality, appearance and comportment. Drawing on motifs of deviant sexuality and symbols of excessive femininity allows dancehall women to express sexual power and affirm their own sexual objectification at the same time. By presenting the

Two women to don's funeral. Photo: The Gleaner File

fleshy female body as unruly and hyper-feminine, dancehall women show femininity to be a masquerade, a kind of mask. The question then arises about how to theorize this masking process and its intent. In European intellectual history, theories of masquerade can be crudely characterized in three ways: first, the feminist tradition, which views masking in terms of patriarchal scopic economy which hides and stifles the true identity of women (Irigaray 1985; Mulvey 1989). Second, the psychoanalytic tradition, which regards the mask of femininity as covering a non-identity, that is, there is no substantive identity to reveal beneath the mask (Riviere 1986; Heath 1986). Both these perspectives view masquerade as absence, superficial, lack and negation. In so doing, they repeat specific assumptions about the relation between surface and depth, inner and outer, masculine and female. In contrast the third, broadly phenomenological, approach holds that identity is constituted *through* the mask itself. On this view {which traces its roots back to Nietzsche}, there is nothing behind the mask, except yet another mask and yet another mode of identity imbricated within another potential surface. In this sense, the layerings that constitute the feminine are neither a hidden essence, nor a concealment of absence; rather, the mask is resolutely dynamic (see Karim Benammar). As one layer yields to another, identity-as-essence gives way to identity-as-performance. In this sense, femininity as a masked identity in dancehall is therefore both a performative and a generative space of being, where the binary opposition between the authentic and inauthentic has no traction.

It is important to remember that dancehall fashion operates within a patriarchal economy which positions women as the object of male desire and control. The extent to which the explicit celebration of women's sexuality challenges the power relations between the sexes is therefore limited. Dancehall fashion should, however, be understood as both an expression of female agency *and* the opportunity for male scopic mastery. For Diane McCaulay, a contributor to Jamaica's newspaper *The Gleaner*, although dancehall fashion offends her feminist sensibility because it reinforces the image of women as sexual objects, she nonetheless sees it as a statement of female strength and agency. She goes further:

> I have to say mostly, though, because I confess to a certain admiration for the outright defiance of dancehall fashion. Further, dancehall clothes cannot be said to promote feminine weakness; on the contrary, dancehall women are

clearly not to be messed with. I am particularly uncomfort-
able with the inordinate interest shown by the authority
figures (usually men) in the way women dress, and the
quantum leap of ascribing national moral decline to the
popularity of a certain type of female fashion. It is a very
short distance from excluding dancehall from festival to
not allowing it at all. Who is going to decide? Apart from
who decides, how are we going to establish what is unsuitable entertainment? What about carnival? Many members
of the middle-class deprecate dancehall as obscene, while
embracing the equally lascivious gyrations of soca.... But
there is clearly no real difference between the display of
skin and the drunkenness of carnival revelry, and the out-
rageousness of dancehall. (*The Gleaner*, 1994)

Dancehall women therefore expose femininity as a performative
and generative construction that calls into question the Euro-centric
and patriarchal separation between the Madonna and the whore, life
and theater, real and appearance. The eroticized exposure of ample,
black female flesh in revealing clothes thereby disrupts the patriarchal
schism between the sexualized female body and the maternal body.
Mary Russo (1986) argues that in Western culture, the exposure of
the female flesh, especially the aging, fat or pregnant body, is seen as
vulgar and grotesque, "making a spectacle of oneself." It is precisely
this kind of female corporeal spectacle that dancehall women seek
to celebrate, as they reject and complicate the dominant reduction of
female identity to a maternal or sexual binarism and the veneration
of the slender ideal. With such self-presentations, the dancehall
woman "puts distance between herself and her observers" (Evans
and Thorton 1991:55). Evans and Thorton argue that this distance
enables women to create "a space in which to manoeuvre and to
determine the meanings of the show" (ibid.). Within a sub-cultural
space that both challenges official discourse and yet nonetheless
reproduces its patriarchal scopic framing, dancehall women have
found a way to intervene through their corporeal styling. Through
sartorial eloquence, dancehall women invite the male gaze only to
fend off scopic capture. This is due to the extravagant, risque style of
their adornment, their dancing skills and the unsmiling, distant look,
which can only be responded to with awed silence by the appreciate
audience—both male and female. Therefore, what appears initially
as sexual vulnerability or availability becomes a form of defensive

armory in which the women assert their own subject position and an unwillingness to be intimidated by the normative pressures of a passive femininity.

Beyond aesthetics and a jamming of the class syntax in Jamaica, a non-European relation to embodiment and corporeal expression is at work in dancehall. Instead of restraint and a self-confining attitude towards the body, dancehall women call upon alternative traditions available within their culture. In this social space, women push the African love of ceremonial pomp to its absolute limit. The desire for full-bodied women in Jamaica is celebrated by the women themselves. Similar to the punk style described by Dick Hebidge and others, dancehall fashion exhumes an iconography of sexual kinkiness "from the boudoir, closet and the pornographic film and placed on the streets where they retained their forbidden connotations" (Hebidge 1979:108).

This sexualized expression is taken to the limit, as bulky black flesh folds over figure-hugging slashed tops to exude an air of self-assurance and intimidation. This bold attitude and stance tempers the sexual offering; the "bad, vulnerable girl" image invoking a bodily confidence is characterized by masculine bravado. Here, we find a psychological corollary to the explicit forms of adornment among dancehall women: a strident confidence that again seeks to oppose and disrupt the chaste respectability of the uptown woman.

SURVIVAL AND IMAGINATION

In addition to the extreme oppositional logic combating class oppression and the uptown normative violence of feminine respectability, dancehall fashion speaks of the edgy conditions in which many women in Jamaican ghettos find themselves. Here, we can introduce the metaphor of cultural energy. When a group is pushed to the limit of existence and marginalized from the normative centers of the production of meaning and cultural symbolization, the expressive energy that seeks output from the group cannot be released smoothly and easily. This energy builds up and seeks an outlet, like a blocked pipe. What often emerges between the gaps is an uncontrollable and uncontainable explosion of creative excessive energy, as desire transduces itself into expression. For dancehall women, the maximal intensity of their clothing, jewelry and hairstyle is the body's response to the existential conditions of their lifeworld: the noise, chaos and volatility that surrounds. The body transforms itself into an ocular symphony, an expressive machine competing via visual overload with the sound booming out of the sound system. The loud screams of their clothing, jewelry and hairstyle is a retort against the gutteral tempo of the music. Sartorial excess emerges as a solution to a limit situation: what Jamaicans call "sufferation."

Women's over-investment in extreme forms of adornment in this culture is therefore an essential aspect of what it means to survive. Against the legacy of plantation slavery, global economic inequality, hetero-patriarchal constraint, the violence of class inequality and religious conservatism, survival entails the transcendence of social death, through an attempt to overcome the horrors and anxieties of daily life. Survival for dancehall women means the attempt to generate meaning and sense in a context of profound meaninglessness and senselessness. In a context where class, patriarchy and color combine to create a unique blend of violence and erasure, survival finally concerns the search for identity, honor and prestige; the attempt to fuse the metaphysical, the spiritual and the existential in a way that allows the urban poor to "imagine an undominated fruition and to live within existing dominations equipped with a determination to do more than survive" (Simone 1989:158).

> Sartorial excess is the way in which transcendence (however temporary) from the harshness of life in the ghetto is imagined and manifested. Female dancehall fans may live under the threat of daily assault and opportunity-denying poverty; they are, however, the best-dressed poor women. Here, there are parallels with other forms of sartorial baroque practices operating as a survival strategy in different cultural settings. The example of Sapeurs mentioned above is a case in point. Like the Sapeurs, for dancehall

women, baroque stylization of the body transcends issue of style towards the fabrication of identities, whereby conspicuous consumption is an aspect of the maintenance of self, Friedman writes, Consumption is a life-and-death struggle for psychic and social survival and it consumes the entire person. If there is a fundamental desperation at the bottom of this activity it is perhaps related to the state of narcissistic non-being generated by a social crisis of self-constitution. (1994:106)

The violence, anxiety and vulnerability of daily life are stitched into the designs of the fabric. The culture of gunning and knifing down opponents that characterizes ghetto life is visually woven into the sheered tops and the motif of bullet holes in jeans. This re-presentation of violence has the effect of both foregrounding existential reality and erasing it at the same time. In their reading of Schiaparelli's 1937 "The Tear Dress," Caroline Evans and Minna Thornton have suggested that by combining the motif of violence with sophisticated high fashion, "Violence and eroticism are simultaneously displayed and made to disappear; beauty is brought to bear on rupture" (1991: 50). Baroque sartorial practice becomes the very means by which daily violence is tamed.

This examination of the issue of adornment in dancehall sheds light on a specific but ingrained contingent relationship between the question of "survival" and that of the imagination, expression and identity. Here, we find a marginalised majority, impoverished, vulnerable and violated in the midst of an urban and spiritual decay inflicted ultimately by the abstract authority of Western capital and Europeanized class ideology. At this point, we are faced with a puzzle that requires something more than the explanation based on the model of cultural energy referred to above: how can this basic subversion of need in favour of an excessive imaginative expression occur in such extreme circumstances? The expectation is quite the opposite: that it is only in the case of material comfort that questions of need and survival can be forgotten, replaced by the bourgeois recourse to worlds of the imagination. In this case, how can this vigorously unique style fabricate itself within such a crisis-ridden space?

To answer this question, we need to turn the assumption that creative excess follows material abundance (and its inverse, that attenuation follows material depletion) on its head. It is clear that instead of emaciated abjection, dancehall women respond to their sociopolitical and economic plight through the multi-textural imprint of enculturated cloth. In a culture where appearance fixes and positions the agent according to a rigid signifying system, dancehall fashion and adornment actively contests and subverts the system of classification itself. Instead of allowing the body to collapse or be rendered mute and inexpressive inside the anxieties of everyday life, through clothing the body is presented as a voluptuous, radiant transcendence of crisis. The "survival" at work in dancehall conforms to the phenomenological logic of the mask: instead of a fixed order of *necessity*—of the basic physical issues of need and demand—necessity is undercut by the performative subversions of desire and expression:

> Among the "people from below" the device of "masking" (in fancy-dress) persists with a vengeance. We still have reason to devise masks to disguise, to create music to affirm, and to assemble dances to celebrate. The ambush of a less than just society under the cover of festive masquerades has been one way of experiencing control, if only a temporary one. Being King or Queen for a day was a way of having a taste for power, even if it was mock power and fleeting... The actual dress is important. For the costume is a mask helping to transform the persona to do wild and uninhibited things—much tulle, dark glasses replacing the old meshwire masks the Jonkonnu characters wore and still do, sequins and costume jewellery, beads, baubles of all kinds, earrings (knobs or droplets), all reminiscent of the pieces of broken mirror on the fancy dress of traditional Jonkonnu! (Nettleford 1995: 16D)

As with the Sapeurs, survival involves an imagination that refuses to be flattened by the forces of negation at work beyond the frontiers of the ghetto. In this way, a truly radical transgression is in operation: dancehall bears no conformity with even the basic existential/social hierarchies of Western normative assumptions. Dancehall women live in such extreme circumstances—of crossfire, acid attacks, rape, spousal abuse, sole caregiver and negation—that fearlessness itself becomes the only mode of survival. It is the extremity of circumstance—the toughness of a life that has been pushed to the limit and is no longer afraid of death or pain—that enables fearlessness to arise. On the one hand, having nothing to fear can lead to senseless acts of violence; with dancehall women, however, it leads to a refusal to acquiesce to the pseudo-transgressive logic of the carnival which Jamaican elites find more agreeable, where a normative order waits to restore normativity behind the mask. It is precisely in this liminal situation that women finally can lose any anxiety over the social perception of their appearance. Instead of the fear of being called a "skettel" or a tart or seen as obscene, the dancehall woman dresses for herself and her community, without care for Jamaican uptown decorum and respectability.

GLOBAL AND LOCAL FLOWS

At first sight, dancehall fashion appears to be strangely dislocated from the rest of society, an orchidacious sub-culture extruded from beyond the rim of mainstream culture. Dancehall adornment springs out like an alien life form complete with its own entirely separate dynamics of existence and taste, like fungi on a tree. Dancehall fashion seems to be the very antithesis of conservative restraint and the concealed style of both official culture and Rastafarian gender coding. It appears to be totally divorced from any cultural or historical context or continuity; springing forth with mutant abundance in a flash of audaciously colored wigs, raucous screech-screaming lamè and sequined tops stretched revealingly across expanses of black flesh. And yet, this dislocated appearance is deceptive. Closer examination of dancehall style reveals deep cross-cultural and historical connections at work in its constitution, demonstrating an attunement with a hybrid array of cultural elements.

In an age of transactional flow of bodies, information, goods, mass media and images, cultural eclecticism has become the only response. Jamaica's proximity to North America, with the more than 60 American TV channels that beam into Jamaican homes, combined with the large number of Jamaicans in the diaspora, means that cultural influence and exchange is the norm. Like their counterparts in the sonic world, dancehall fashion also favors sampling, cutting and mixing in order to give birth to something different and distinct. Dancehall women have raided the global wardrobe and given it local texture. Odd and incongruous materials, imageries, accessories and patterns are combined to produce dizzying and dazzling layers of material, texture and form. Plastic, lurex, polyester, lycra, nylon are combined with leather, silk, organza lace, velvet, brocade. Late 1960s hot pants ride higher into the buttocks to reveal more than its original ever did; English granny purple-rinse hair styles become an unrecognizable, chromatic sculptured coiffure on youthful bodies; the "cut up" and bondage straps of punk were cleaned up and re-emerged as the "air-conditioned linen summer wear" designed by Sandra Campbell; Vivienne Westwood's 1976 Bondage Collection was hungrily retrieved and stripped of its Nazi associations. Punk girls' fishnet tights became the now classic mesh string vest. Fake chunky 22-carat gold evoked the tradition of goldsmith's art that flourished in the former Gold Coast of West Africa, Ghana, denuded of its royalty and hierarchy and reduced to a kitschy repetition of the original. The bridal nose ring linked to the ear by a fine delicate chain connects dancehall women to a tradition of Bollywood-esque ostentatious display. All these motifs are emblematic of the way in which dancehall women have absorbed and adapted global fashions, goods and images, and inflected them with new meanings that have made them refreshingly and uniquely Jamaican. This is what Hudita Nura Mustafa, in her account of changing fashion in Senegal, has termed the "sartorial ecumenc." doing the rounds in Jamaica around the same time? Are Versace's gold-encrusted jeans an original which the dancehall fashion aficionados imitated? The answer to these questions is impossible to resolve. The line of influence of any cultural artifact is often not obvious or explicit; in dancehall the line is a spaghetti sprawl of loops and curves. Origins and originals are the stuff of bourgeois distinction-making and intellectual property rights—elements with no meaning in a world where the distinction between fake and real has little purchase. Rather, the incessant appropriation, re-appropriation and expropriation of global/local exchanges reveal the intrinsic mutuality of cultural artifacts and patterns of expression. Through dancehall's sartorial borrowing and transfer, fashion is perpetually caught up within a dynamic of differential repetition that has multiple origins.

In this article, I have argued that dancehall fashion and corporeal stylization show how women inured by life in the urban ghettos of Kingston interpret their life world, inflect it with meaning and recycle the different cultural artifacts circulating within a global economy of sartorial signs. Beneath interpretation and a semiotic analysis, however, I have indicated that dancehall styling is ultimately a question of survival; an excessively imaginative response to the class, race and gender-based normative violence of the hegemonic morality of the uptown elite. I hope to have encouraged an appreciation of the ways in which, prior to speech and lyricization, women in dancehall culture exhibit an expressive styling that ultimately should be characterized as a defiant performance of generative identity in the midst of perpetual existential crisis.

Notes

1.   This is not to deny that women do also contribute to the production of music as is evidence with the prominence of female dancehall DJs such as Lady Saw, etc. However, through sartorial practices a larger group of women can participate and contribute to dancehall culture. Fashion and adornment thus becomes a democratic space which allows different categories of women to participate in the production of symbols and cultural meaning.

2.   Not all female participants in the culture dress so flamboyantly, however. Dancehall women therefore refer to those women who spend a considerable amount of their time and resources attending the dancehall events, drawing attention to themselves at any event through their fashion style which stands them out from the rest of the crowd. It is these women known as dancehall "divas" or "donnets" who have aroused interest, fascination and vilification all at once who are the central subject of this article.

3.   During the fieldwork for this research it became clear that some of the subversive and transformative potential I attributed to dancehall women were not always explicitly shared by dancehall women. Many pointed out that they had not intentionally set out to challenge hegemonicstructures. Rather, they were simply dressing for themselves and the dancehall space provided them with the opportunity to express themselves. The gap between how participants understand and explain their action and my own interpretation of their action in no way detracts from the argument. It is important to remember that Jamaican women have a long history of resisting oppressive regimes and articulating their existential positioning using a variety of media. I locate dancehall women's sartorial expression as a continuation of this history of resistance and cultural production.

4.   It will be true to say that Dancehall music and culture has now become mainstream, but its energy, creativity and reproduction continue to be drawn from the socio-political and economic realities of the marginalized urban poor. Despite being a major Jamaican

cultural export, Dancehall still occupies an ambivalent place within Jamaican elite cultural imaginary.

5. In Let those Monkeys Out, Dancehall DJ Lexxus sings about the use of bleaching cream by dark-skinned women to acquire the browning effect, and their increasing use of fattening pills used in industrial chicken-farming to acquire a ample body is so valorized in the culture. A verse from the lyrics goes,

Me hear some o' them nuh stop take the fowl pill '

MmMm, so me know she them gyal there skill

A take the fowl pill just to impress Phil

When you see them you fi shout, 'Dill dill'.

REFERENCES

Austin-Broos, Diana J. (1994). "Race/Class: Jamaica's Discourse of Heritable Identity." *New West Indian Guide* 68(3/4).

Banks, Ingrid. 2000. *Hair Matters: Beauty, Power, and Black Consciousness,* New York: New York University Press.

Barnes, Natasha B. 1999. "Face of the Nation: Race, Nationalisms, and Identities in Jamaican Beauty Pageants." In Consuelo Lopez Spring-field (ed.) *Daughters of Caliban: Caribbean Women in the Twentieth Century.* Bloomington: Indiana University Press.

Barrow, Steve, and Dalton, Peter. 1997. *Reggae: The Rough Guide.* London: Rough Guides.

Benammar, Karim. "The Mask and Self in Contemporary Japanese Philosophy." Unpublished paper, http://www.plethora.nl/benammar-maskandself.htm, accessed January 2002.

Bilby, Kenneth. 1997. "From "Jamaica."' In Chris Potash (ed.) *Reggae, Rasta, Revolution; Jamaican Music From Ska to Dub.* London: Schirmer Books.

Bourdieu, Pierre. 1977. *Outline of A Theory of Practice.* Trans. R. Nice. Cambridge: Cambridge University Press.

Breward, C. 1994. *The Culture of Fashion.* Manchester: Manchester University Press.

Chude-Sokei, Louis. 1997. "Postnationalist Geographies: Rasta, Ragga, and Reinventing Africa." In Chris Potash {ed.} *Reggae, Rasta, Revolution: Jamaican Music From Ska to Dub.* London: Schirmer Books.

Connerton, Paul. 1989. *How Societies Remember.* Cambridge: Cambridge University Press.

Cooper, Carolyn. 1993a. *Noises in the Blood: Gender and the "Vulgar" Body of Jamaican Popular Culture.* London: Macmillan.

_____ 1993. "Emergency (Un)Dress." 1993b. *Lifestyle* 27, July/August.

Coopeer, Carolyn. 2000. "Virginity Revamped: Representations of Female Sexuality in the Lyrics of Bob Marley and Shabba Ranks." In Kwesi Owusu (ed.) *Black British Culture and Society: A Reader.* London and New York: Routledge, pp. 347-57.

D'Elia, Susan. 2002. "Women's Fashion in Jamaican Dancehalls." http://debate, uvm.edu/dreadlibrary/delia02. html

Douglas, Lisa. 1992. *The Power of Sentiment: Love, Hierarchy and the Jamaican Family Elite.* Boulder: Westview Press.

Evans, Caroline and Thornton, Minna. 1991. "Fashion, Representation, Femininity." *Feminist Review* Summer 38: 49-61.

Fanon, André. 1988. "Is Dance Hall Music a Creative Force?" *The Gleaner* (Jamaica), 29 November.

Friedman, Jonathan. 1994. *Cultural Identity and Global Process.* London: Sage.

Hall, Stuart. 1992. "What is this 'Black' in Black Popular Culture?" In G. Dent (ed.) *Black Popular Culture.* Seattle: Bay Press.

Heath, Steven. 1986. "Joan Riviere and the Masquerade." In V. Burgin,

J. Donald and C. Kaplan (eds.) *Formations of Fantasy.* London: Methuen.

Hebidge, Dick. 1979. *Subculture: The Meaning of Style.* London: Methuen.

Hurston, Zora Neal. 1933. "Characteristics of Negro Expression." In Gena Dagel Caponi (ed.). 1999. *Signifyin(g), Sanctifyin', & Slam Dunking: A Reader in African American Expressive Culture.* Amherst: University of Massachusetts Press.

Irigaray, Luce. 1985. *Speculum of the Other Woman.* Trans. G. C. Gill. Icatha, NY: Cornell University Press.

Jahn, Brian and Tom Weber. 1992. *Reggae Island: Music in the Digital Age.* Kingston: Kingston Publishers.

McCaulay, Diane. 1994. "Dancehall and Choice." *The Gleaner* (Jamaica) 12 April.

McDonald, Paulette. 1993. "The Dancehall Revisited." *Lifestyle* 27, July/August.

Mercer, Kobena. 1987. "Black Hair/style Politics." *New Formation* 3, Winter.

Mintz, Sidney \V. and Richard Price. 1976. *The Birth of African-American Cultures; Anthropological Perspective.* Boston: Beacon.

Muivey, Laura. 1989. "Visual Pleasure and Narrative Cinema." Laura Mulvey *Visual and Other Pleasures,* London: Macmillan.

Mustafa, Hudita Nura 1998. 'Sartorial Ecumenes: African Styles in a Social and Economic Context." In *The Art of African Fashion.* The Netherlands and Eritrea/USA: Prince Claus Fund and African World Press, New Jersey.

Nettleford, Rex. 1994. 'Dance-hall, Part of the Jamaican Heritage?, *The Gleaner* 22 July.

Nettleford, Rex. 1995. "From Jonkonnu to Dancehall." *The Gleaner.* (date illegible from original)

Polhemus, Ted. 1988. *Bodystyles.* Luton: Lennard.

_____ 1994. *Streetstyle.* London: Thames & Hudson.

Riviere, J. 1986. "Womanliness as Masquerade." In V. Burgin,J. Donald and C. Kaplan (eds.) *Formations of Fantasy.* London: Methuen.

Russo, Mary. 1986. "Female Grotesques: Carnival and Theory." In Teresa De Leureatis (ed.) *Feminist Studies/Critical Theory.* Blooming-ton: Indiana University Press.

Simone, Timothy Abdoumaliq. 1989. *About Face: Pace in Postmodern America.* New York: Automedia.

Stolzoff, Norman C. 2000. *Wake the Town and Tell the People: Dancehall Culture in Jamaica.* Durham, NC: Duke University Press.

Tseëlon, E. 1997. *The Masque of Femininity.* London: Sage.

CHAPTER ELEVEN

Desire through the Archipelago

THOMAS GLAVE

As a child of Jamaican immigrants to the United States, and one who regularly spent extended time with family in Jamaica, I couldn't know how much, over the accumulating years, a book like this would someday mean to me. It is both fair and accurate to say that this anthology—this gathering, as it is tided, which makes its own contribution to an ever increasing conversation—is a book that I and others have been waiting for and have wanted for all our lives. It is in no way an exaggeration to say that this gathering originated as an idea born out of the most extreme longing: the desire to know finally, and with complete certainty, that a book such as this one actually existed and could exist. Could exist in spite of thundering condemnation from Christian fundamentalist ministers and, from those in churches, mosques, and other places, sidelong disapproving—and sometimes baleful—glances. Could exist despite proscriptions, banishments, ostracisms, and, in more than a few cases, extreme violence.[1] Could exist: a book like this that, though some—including a number of our most renowned, if not always most generous, Caribbean minds—might wish to ignore or dismiss it, none would ultimately be able to deny or wish away. Numerous writers whose work I have long admired have consistently expressed a sentiment similar to the one that motivated this collection, also centered in desire; they wrote the books they did, they said, because they wished deeply to read those texts, texts that up until the moment of that writer's yearning had not been written. Those writers wanted to know that the stories they committed to paper actually existed— indeed, could exist. Exist for the passing along, the making known from consciousness to uttered word to the next watchful, waiting eye.

I wouldn't be truthful if I didn't admit that this gathering also owes its genesis to the most painful, inexpressible loneliness: one that, as the years passed me from childhood to adolescence to adulthood, grew increasingly agonizing, made more so by the sorts of intransigent silences I fervently hope the voices in these pages will unsettle. That loneliness stretched through every hour of the silent and silenced wondering one couldn't possibly dare, when young, to ever express out loud to anyone else, or—perhaps most dangerously, terrifyingly—to oneself. Wondering: could there somewhere "out there" be others like me? Wherever "out there" was, I knew it to be a place beyond the supposed safety of family (which was not always so safe)—a place where I was despised, even hated, by those who ranted that hatred publicly and shouted it joyously in church. Were there others in Jamaica and other parts of the Caribbean who also pondered "unspeakable" things and thought, in the depths of their own silences, "unmentionable" thoughts? Were there others immobilized and cowed by silence and mired in the same shame that colluded with that silence and its creeping shadows? Shadows that, like the heavily weighted silence, invariably led to rage, self-loathing and desperation. Desperation backed the thoughts: *I hope someday I can get out of here. I hope I survive long enough not to walk, for the last time, with open eyes and open mouth into the swirling sea. I hope that star-apple trees and bougainvillea blooms never learn my secrets—no, not any of them—and that the hummingbird that adores scarlet hibiscus never penetrates my dreams. I hope that this silence doesn't kill me or make me kill myself, because* (some of us thought and continue to think) *it doesn't seem as though I can possibly be myself, my fullest truest self, the self that everyone would love to know and hug and laugh with, greet with open hands and arms, if I remain here... remain here.* Some of us, myself included, thought those things. Some of us, in spite of those thoughts, did not leave the Caribbean, and will not leave it still. Although some opted to stay, many had—and continue to have—no choice. Green cards are not always easy to secure, nor are visas, for those who need them most. And families, families often are in need of help. The desire for Home or "home" often abides in the traveler. Our history—perhaps especially the history we find most intolerable to remember, that began for some of us with harsh voyages across the sea, but never ended there—is about nothing if not movement, memory. Dis-placement. This book began, crept to its knees, and shakily, then more assuredly, began to walk through all of that and more.

Walking, though slowly, and increasingly aware of its flesh taking shape, the book began to ask questions—pester, even.

Through all the hours of its more groggy, wobbly-kneed state, it wanted to know the same things I did: things about the people like me whom I longed so much to know (and who, it seemed, were so often impossible to find) in Jamaica, Martinique, St. Lucia, Trinidad. People whose eyes would say something recognizable, friendly even, in the Dominican Republic, Sint Maarten, and Guyane Francaise. People "out there" who also gazed across that water that simultaneously divided and united us all and who dreamed—yearned their way through those emotions and all that desire: women for women, men for men, women and men for women-men. That erotic-emotional desire for people of our own gender that it seemed no one—not anyone at all—ever spoke about, much less wished to hear about unless in the realm of "scandal" and "disgrace."

Those people, yes. The people of "out there" and, as I would later learn, nearer. But where did they live? How did they dream?

What colors would their stories have taken, if one could have seen them? Read them? How would they have sounded?

Who, through love and desire, loneliness and pleasure, were we hoping to be, imagining ourselves to be, and becoming through our many Antilles? Our Antilles of Creole and English, Kréyol and French, and Spanish, Papia-mentu, Dutch. Our many Caribbeans of cricket and *béisbol*, soccer and *fútbol*, and Carnival, jouvert, crop-over. Our islands and memories of *zonas colo-niales, ciudades viejas,* a very well known Soufriére, and Basse-Terre, Grand-Terre —all traversed by mountains, rivers, ghosts, *diablesses,* duppies, sou-couyants, jumbies, and more.

And all of our music! Zouk. Salsa. Rara. Bouyon. Reggae. Rumba. Calypso. Ska. Son. Tumba. Biguine. Bomba. Dancehall. Soca. Merengue. And still so much more.

What would it be like to attend—to truly hear, for once—the many conversations that we have had with each other and still need so very much to have? What would it be like to listen to and now, by way of a gathering of voices like this one, actually observe those conversations between ourselves?

What would it be like to bear witness to the conversations between *mati,* women-loving women, in Suriname and their distant sisters and brothers in St. Eustatius, Saba, and Bonaire?

But then close your eyes, a certain tree tells you, in order to hear better the words of those two: two women together, or two men, or two women-men or men-women. They are in Haiti—yes, in Jérémie, this time. There, in that near-darkness, overlooking the sighing shoals and what remains of the ever-entangled mangroves, they are holding each other, those two. Holding, caressing each other's most secret parts, and whispering things into each other's neck—things, the tree says, that their brethren and sistren in Fort-de-France, Marie-Galante, and even all the way over in Cayenne, Guyane, would like to hear. We too would like to hear it, the tree tells us. (In its own, old way, it is one of us, that tree. Of course. It always has been.) We would like to hear all of it, for the expansion of our imaginations and our very lives—our survival. It is so very important for us each to know that we are not, no matter what anyone tells us, throughout the archipelago and beyond it, alone.

INCIPIENT CHALLENGES: NOTES ON "NARROW" AND "A SLICE OF A SLICE"

When I first conceived of this gathering as a viable project, I was taken aback—though in the increasingly corporate-minded publishing world, sadly not entirely surprised—by the opinion of a few myopic editors and one remarkably uninsightful literary agent, who felt that an anthology like this one would be a "narrow" book: "a slice of a slice," as one extremely market-minded editor put it; that is, the (non-Latino) Caribbean "market" as perceived by some continental North American editors was a "slice" and a lesbian and gay text within that category an even more selective "slice." (Prompting the question of who or what was looking down from somewhere, holding the determining slicing knife?)

After hearing this editor's words—an assessment that instantly struck me as informed mostly by ignorance and prejudice —I immediately lost faith in her judgment. Fortunately for this book's ultimate existence, I eventually made contact with a more broadminded editor—one in possession of the imagination and generous vision the naysayers lacked, who not only decided to take on this project but from the very start voiced his unstinting support for it and belief in its importance.

This gathering's blessed fortune in that regard notwithstanding, I remained deeply troubled by those people's summary dismissal of

Caribbean lesbian and gay lives. How, I wondered, could they feel so confident in their arrogance? Confident enough to use the dismissive word "narrow" about our very complex and, to them, unknown and unseen lives? What, I wondered, did they in their continental North American worlds—contexts rife with spurious images and conjurings of the Caribbean as a fetishized "paradise" for tourists—really know about our lives ? The Caribbean, as packaged globally for tourists, purposely obscures quotidian (and often poor) Antil-lean lives—existences rendered as of scant importance beyond obsequious servitude in the generally consumer-directed packaging. Lesbian and gay lives generally do not enter into this truncated representation at all, unless they surface in some momentary wink of sexual tourism. The naysayers, editors and agent, were not Caribbean themselves. None among them, as far as I knew, self-identified as lesbian or gay (although I remain uncertain even today that an editor from either group would have seen things any differently) . If they considered a book containing texts about our lives "narrow," what would they have considered the opposite? An anthology containing work by U.S., Canadian, or European heterosexuals?

I recount these challenges in order to bring to light the very difficult, *un*welcoming context out of which this anthology fought—and was required to fight—for its right to exist; to point out how, in that context, many in the "global north" or "first world" think little of those in the "global south" or "third world" and frequently have no problem in saying so; and to briefly delineate the reprehensible attitudes (narrowness, actually) of some publishers (including, in this case, more than one very respectable university press) who retain the power to decide which ideas achieve visibility under their aegis and which do not—including books by and about people whose life (and death) experiences are continually rendered invisible, or silent, or both.

ORIGINS: GATHERING

Sometime during the late winter to early spring of 2002, when I realized, startled, that my longstanding wish to see and hear Caribbean lesbian and gay voices in conversation could actually be a book, I began to speak with and write to people who I very much hoped might be interested. Whatever else occurred, I knew from the beginning that I wanted this project to move, as much as possible, as a grassroots effort: one that reached out to people across and through the varying Caribbean communities I had known and of which I had been a part, as well as those unknown to me, but viable. These communities included women-loving women and men-loving men, all of whom were in one way or another aware of and affected by the issues that have always weighed heavily upon us: racism, xenophobia, sexual and gender violence, economic stresses, the need for voluntary or forced migration (as in the case of political asylum seekers), the erasing or silencing of ourselves (and, in some cases, the murdering of ourselves) because of our sexualities, and more. The book I yearned to read was one that would disclose *how* we wrote and continued to write. And why. And where. That book would suggest questions like: writing and thinking as we did, did we remain in the Caribbean? Was that decision ours to make, or someone else's? Could we remain in that place—the place we might have called home—as the people we were and continued to become? That is, as our fullest, most realized selves? Would those places we thought of as home provide a welcome for us whomever we turned out to be and whatever we decided to do or say in artistic, creative, and intellectual response to those places and to ourselves as people ancestrally, imaginatively, and culturally part of them? For whatever else happened, we were part of those places and their people (our people) in just about every way. We remained connected to them, even as—perhaps especially as—members of an enormous diaspora. What effect would the decisions we made and the actions we took—artistically, intellectually, politically—extend not only to our literal movements, but also to our internal ones in the realm of our imaginations where, as people involved with language and words, we think in each of our very different locales about the ways in which we might not only express ourselves, but also challenge ourselves and others? How might a journey to our particular diaspora, a settling in that diaspora, and a return to "home" from that "foreign" place affect or not affect our choice of literary topic and the way we chose to imagine it, shape it, inhabit it?

As same-gender-interested people, what was and will be our relationship to the colonial powers that impressed their languages upon us and convinced us that their tongue was not only the "correct" one, but the word of the Almighty (Christian) God? Such questions arose for me and for many other writers featured here as the collection developed.

It was not my intention to focus on any one piece of writing in this gathering, but partly in response to some of the questions above and others to come, and also for additional amplification, I wish to draw attention to the rarely seen, little known essay "Of Generators and Survival: Hugo Letter," by the U.S. born, Caribbean descended author and activist Audre Lorde. This work, though written by someone who constantly, energetically identified herself as a lesbian—indeed, as a black, feminist lesbian —concerns itself not at all with matters explicitly lesbian or gay, but with the devastation wreaked by a Caribbean hurricane in the "developing" world and its lingering, dire effects of poverty, colonialism, racism, and imperialism: realities beneath which all Caribbean peoples, especially the poorest, suffer, and which a natural disaster, in its damage of infrastructures and vital resources, horrendously highlights. In this regard, Lorde's essay strikes me as bearing particular relevance to all Antilles-connected people, including those of us—perhaps especially those of us—erotically and romantically inclined toward those of our own gender. "Of Generators and Survival," as all of Lorde's lifelong writing and activism, reveals to us that even in the most integrated, humane world possible, none of us can afford to indulge in single-focus politics, irrespective of how we name ourselves or are named by those who have the power to do so. For the majority of us, the practice of single-focus politics is essentially always impossible, given that virtually none of us on this planet occupies a single-focus existence. Lorde's blackness, womanness, and lesbianness, though barely discussed in political terms in "Of Generators and Survival" nonetheless occupy center stage in the black, female, lesbian body she physically moved to help her fellow citizens in St. Croix after Hurricane Hugo. Like Lorde's essay, the works in this book demonstrate their authors' concern with the varying issues that impact on the lives of Antilles-connected people. These issues, which originated in our labyrinthine history, still have neither been completely confronted nor fully understood by many Caribbean peoples; by the empires that conquered, enslaved, and transformed us; or by the tourists and wider public who think of us (when they think of us at all) as people of "the islands."

With this book's emergence, I anticipate that the voices collected herein will provoke radically different interpretations of all of these considerations and of innumerable others. Above all, I hope that each reader will journey through this gathering and garner much pleasure along the way from the questions and possibilities that not only will loom but also will deepen, widen.

GATHERING/LANGUAGE

Language itself posed one of the most vexing questions at the center of this project's development, principally in the consideration of a tide. What title might appeal to and invite into the collection as many as possible, especially those most in need of it, while representing as generously as possible all our vastly different ways of being Caribbean, as well as our reflections on our experiences as same-gender-interested people? What words could evoke and, if possible, encapsulate the nuances, complexities, and powers of these gathered works?

At first, quite ethnocentrically and with the blithe naïveté that some ethnocentrism brings, I considered a tide that would have used Jamaican Creole words. If any tongue were to represent the collection, I reasoned, surely a Creole or "patois" language would be the most appropriate offering, given the often rancorous debates among Caribbean people about Creole languages and the outright contempt frequently expressed for those tongues as dialects and "broken" or "bastardized" versions of a more "proper" language. I thought that I would choose a Creole language that, like many of the people who routinely used it, was regularly overlooked, disdained, trivialized, and regarded with irritated amusement and condescension, if not outright disgust: a tongue developed out of the mourns and memories of slaves and others whose presence had made possible the Caribbean of our lives today. Fortunately, I quickly realized that given this gathering's pan-Caribbeanness, my working as a Jamaican "supremacist" in this regard would be an inexcusable act of nationalism thickened by naive nostalgia.

That said, not long afterward, I threw up my hands in resignation at the realization that, for a variety of reasons, I could see no way around the decision—an editorial decision, entirely my own—to make English the book's representational language, with all of the texts originally written in other languages produced in these pages in translation. In respect to my Caribbean sistren and brethren whose first tongue is not English, including those in

English-speaking Antillean nations, I remain uncomfortably aware of the difficult and problematic realities surrounding my decision, especially given the English language's controlling force in the world at large and its potential to erase, here and elsewhere, the tongues of those who do not (and do not necessarily have any wish to) write, speak, think, dream, or imagine in English. For many reasons, this particular language dilemma was one that I could finally bring to no truly satisfactory conclusion. Yet with the supreme importance of our diverse tongues acknowledged and grasped, I am glad—at least in this incipient moment, though some of us speak here through the scrim of translation—that we are in this gathering truly present: speaking and listening to each other as the neighbors we are and have always been. It is this speaking and listening that I am certain will lead us closer to what one poet long ago imagined out loud as the dream of a common language; as we acknowledge the fierce reality that (to paraphrase the title of another poem, one featured in this gathering) no language—not any language—can ever be neutral.[2]

For me, this collection's title remains partly vexing for its use of the words *lesbian* and *gay*—for what exactly, I wondered along the way and wonder ever more today, is—can be—"lesbian and gay writing"? Is it writing with "lesbian" or "gay" (or "queer"?) content? Writing by those who consider themselves to be lesbian or gay? By those who consider themselves to be simply not heterosexual? But there are several individuals whose work is featured here who would most definitely not identify themselves as lesbian, gay, bisexual, *zami*, "that way," *asi*, "so," or—as presently heard among some in the English-speaking areas of North America and Europe—queer. What of those who write about erotic and romantic same-gender interests, but who would refuse to consider naming in any "lesbo" or "homo" way their erotic inclinations and ideals—those who might say about themselves, if they said anything at all, that they were women interested in other women, men focused on other men, and leave it at that? How, if only for the purposes of a project like this one, to define or place their work? I use the words "a project like this one" not at all lightly. For if anything, the works in this book are deeply concerned with, rooted in, the efficacies and intricacies of a variety of different languages and words that have historically been forbidden, taboo. If some of us choose to name ourselves with the more defining (or at least, in many quarters, familiar) words *lesbian* and *gay*, and speak of our writing in such terms, we face the challenge put forth by those whose self-naming desires differ from ours of bumping straight up against a language that, like most if not all languages, simply will not, cannot, serve nor satisfy everyone. Yet this frustrating fact has a potentially productive conflict at its center and can make possible for all of us new, greatly unsettling journeys. At the very least, it illustrates our enormous diversity. Contrary to the opinion of some, we are not—cannot possibly be—all the same.

And so, to complicate matters further, what exactly is—can be —"Caribbean lesbian and gay writing"?

The words *lesbian* and *gay* in the tide lost this gathering at least two writers: one a woman of Caribbean background who, while erotically interested in other women, has long refused to call herself a lesbian and, as she told me, wished not to have her work involved with any text that would categorize her writing as either lesbian in content or as authored by a lesbian. In the other instance, the word *gay* in the title gave pause to the executors of the estate of Severo Sarduy, the marvelous Cuban writer whose work I have long admired for, among other things, its formidable experimentation with form and content. I wished very much to include his work here, and would have done so but for objections raised by those who, now overseeing his work, did not wish any of it to be associated with a gay volume. Such objections—by the woman writer mentioned, and others—are real, enduring, and, for an editor working on a project like this one, both painful and frustrating, even enraging; yet the very powerful unwillingness of some people to be associated with anything lesbian, gay, or "queer" illustrates dramatically and often sadly the very great need for this book. If it is impossible to know how many people might steer clear of this collection because of its subject matter, we know that it is simultaneously possible to dream, in the most lustrous colors imaginable, of all those who will be drawn to these pages precisely *because of* their contents.[3] Such dreams bring great pleasure. All pitfalls and initial concerns considered, I am finally very pleased with the title, and especially with its claiming of the Caribbean as "our" place—for in whichever Caribbean we originate (including those we ourselves invent, reinvent, and re-imagine), it all remains ours, irrespective of what those who would rather pretend that we do not exist or who actually would wish us harm have to say.

GATHERING, BEGINNING: OUR CARIBBEAN

It is a great pleasure to present this book to readers—readers everywhere and, I hope, as many in the Caribbean and of Caribbean background as possible. For all of us, wherever we may be and however we may ultimately decide to identify ourselves and pursue our lives, an ending of silences and invisibilities begins here. In the pages that follow, the conversations that began long ago in imagining and desire continue. For as you surely know, and as I have learned and am still learning, the coming together that is this one, and others like it, widens, deepens, and—as far as I can see—has no end.

No end, like the sea: that sea not far from which this all began. Its water still stretches a broad, restless belly out to the sky; just now, the sun meets it, gently wetting first its chin in those waves, then its wide brow; palms flanking the shore murmur as seabirds dip, scolding but more weary at the day's end; cane stalks rustle in the breeze, insects wither, and the evening's peepers commence their choral chants. The water continues to gaze calmly, as it has always enjoyed doing, at the light-speckled shorelines suddenly retreating beneath dusk's quiet step. Quite soon the moon, lumbering up over those hills in the east, will have something to say about all this. But for now, over all that glistening water and beneath it, gathered voices are rising. Rising as they call *Now*, as they call *Here*. As they utter, so softly, *Listen*.

THOMAS GLAVE

Kingston, Jamaica

June, 2006

NOTES

1. By "extreme violence" I refer to the violence and rage enacted against same-gender-interested people in a place which I know well, vividly remember, and to which I continue to return: Jamaica. As an additional note against the cruelty and indifference of forgetting, coupled with the desire to avoid demonizing any locale, I also wish to make clear that extreme violence continues to be visited upon same-gender-interested people throughout the entire world, including nations whose populations consider themselves "first world" and "developed."

2. "The dream of a common language" comes from Adrienne Rich's poetry collection of the same tide; "no language... can ever be neutral" borrows, in paraphrase, from Dionne Brand's poetry collection *No Language is Neutral*.

3. As I acknowledge the joy that this anthology's publication brings me and the pleasure I know it will bring others, I also wish to state here my deep regret in not being able, for a variety of reasons, to include so many other writers' work that really should have appeared here, and which will perhaps loom in future anthologies that will continue the critical work this collection has begun. Foremost among those authors is the magnificent Cuban novelist Jose Lezama Lima, without question one of the Caribbean's and Latin America's—and the world's—greatest literary artists. His novel *Paradiso* remains' a supreme masterpiece, appropriately revered by those fortunate enough to have encountered it. In spite of his absence here, I would like to think that Lezama Lima's towering presence—the fierce intelligence, style, and precision manifest in all his work—moves through this book, even if his actual prose does not.

CHAPTER TWELVE

Bullers" and "Battymen": Contesting Homophobia in Black Popular Culture and Contemporary Caribbean Literature

TIMOTHY S. CHIN

The recent controversy surrounding Buju Banton, the Jamaican dancehall "don"—which, like so many contemporary debates about race, gender, and sexuality, has been played out in the theater of popular culture—demonstrates the high ideological stakes as well as the discursive limits that determine current discussions of gay and lesbian sexuality and Caribbean culture. Occasioned by the circulation over North American airwaves of Banton's popular dancehall tune "Boom Bye Bye," the controversy provides a prime example of the cross-cultural conflicts and contradictions that are often generated by the increasingly globalized markets of the culture industry. The debate, as it was staged in the pages of the popular press *(New York Post, Village Voice)* and periodicals associated with the music industry *(VIBE, Billboard)*, concerned the alleged homophobia displayed in the lyrics of Banton's song.[1] According to an article that appeared in the *Village Voice*, two groups—GLAAD (Gay and Lesbian Alliance Against Defamation) and GMAD (Gay Men of African Descent)—joined forces in 1992 to "decode Buju Banton's bullet-riddled patois" and "embarked on a media campaign to have 'Boom Bye Bye' removed from the playlists of radio stations WBLS and

264 | CHAPTER TWELVE

wrks." Peter Noel and Robert Marriot, the co-authors of the *Village Voice* article, applauded GLAAD for boldly defining the meaning of *diversity* and *tolerance* for Banton. Insisting on a literal reading of Banton's lyrics, Noel and Marriot state that the song "advocates the execution of gay men" and, consequently, reflects the especially virulent forms of homophobia that are rampant in Caribbean culture generally and Jamaican culture specifically.[2]

Interestingly enough, the critics who have—to varying degrees—defended Banton also tend to rely primarily on culturally based arguments. However, these critics typically assert that Banton's lyrics should be understood metaphorically and that metropolitan critics have therefore misread both Banton's song and the "indigenous" culture from which it springs. For example, in a piece written for *VIBE,* Joan Morgan criticizes certain North American reviewers for their "ignorance of Jamaican street culture" and their inability or unwillingness to "grasp the metaphoric richness of Jamaican patois."[3] In addition, Morgan contends that Buju Banton's refusal to apologize for "Boom Bye Bye" "makes the most sense" given his first and ultimate commitment to the "hardcore dancehall audience" to whom Banton owes his success. According to Morgan, Banton's loyalty to this (cultural) constituency has been rewarded—unlike what is conversely seen as Shabba Ranks's capitulation to the powers that be—with an even greater adulation from his "true" fans.[4]

Carolyn Cooper, a well known Jamaican literary and cultural critic, has likewise insisted that Banton's gun is essentially a "lyrical" one that is meant to illustrate "the function of metaphor and role play in contemporary Jamaican dancehall culture."[5] Consequently, Cooper argues that critics who are unfamiliar with the metaphorical qualities of the Jamaican vernacular have misread Banton's song by taking his words all too literally: "Thus, taken out of context, the popular Jamaican Creole declaration, 'aal bati-man fi ded,' may be misunderstood as an unequivocal, literal death-sentence: 'all homosexuals must die.'" In contrast, Cooper suggests that Banton's "lyrical gun" should be understood primarily as a "symbolic penis" and, therefore, "in the final analysis, the song can be seen as a symbolic celebration of the vaunted potency of heterosexual men who know how to use their lyrical gun to satisfy their women."[6]

Although critics like Cooper and Morgan have rightfully exposed the ethnocentrism that typically informs dominant accounts of the controversy—which often suggest, for example, that North American culture is more advanced and therefore less homophobic than its Caribbean counterpart—their arguments, nevertheless, tend to reinforce a notion of culture that relies on certain fixed oppositions between native and foreign, indigenous and metropolitan, us and them, and so on. Even if we concede that their arguments do not seek "to legitimate homophobia on so-called cultural grounds," as one response to Morgan's *VIBE* piece alleges,[7] these critics have nevertheless missed a crucial opportunity to challenge the deeply rooted homophobia that is unmistakably reflected in Banton's lyrics and that, more importantly, pervades Caribbean societies, as it does most third and first world cultures. In contrast to the reactive or defensive postures implied by such arguments, it is necessary—especially given the complex ideological issues currently surrounding the question of black cultural production—to formulate modes of cultural criticism that can account for the differences within as well as between cultures. In addition, our contemporary situation calls for a cultural politics that can critique as well as affirm—a politics that recognizes, in other words, the heterogeneous and contradictory (as opposed to homogeneous and monolithic) nature of all cultural formations. In the words of Jamaican anthropologist Charles V. Carnegie, "Even as we seek to restore 'indigenous knowledge' systems, we must simultaneously seek to sharpen an 'indigenous' criticism."[8]

Despite the limitations that currently define the terms in which the debate has been carried out, the Banton controversy—as Cooper ironically notes—nevertheless opens a critical space for talking about questions of gay and lesbian sexuality and homophobia as they pertain to Caribbean culture.[9] Using this critical space as a point of departure, I would like to continue and extend this dialogue by exploring the representation of gay and lesbian sexuality in contemporary Anglophone Caribbean narratives. Such an exploration implicitly assumes that the texts in question inevitably reflect and, indeed, participate in (by reinforcing or contesting) the sexual ideologies that pervade the wider culture.

If the Buju Banton controversy represents a manifestation of how such questions have recently erupted in the realm of the popular, Caribbean literary production has traditionally maintained a conspicuous silence around issues of gay and lesbian sexuality. In this case, the absence of representation is perhaps the most telling factor, especially when we consider the earlier decades of literary activity. Nevertheless, there are writers—like Claude McKay and Paule Marshall, for instance—for whom gay and lesbian sexuality or "homosexuality" remains an important subtextual issue, and one that is intimately and inextricably intertwined with other, more explicit narrative preoccupations. In addition, there are more recent writers—emerging particularly within the last three decades—who have broken the taboo that has previously surrounded the question of gay and lesbian sexuality and homophobia in Caribbean culture. These writers have vigorously challenged the patriarchal and heterosexual ideologies that have resulted in the marginalization of women and gay men at the same time that they have continued to expose the social and political structures that serve to perpetuate the region's colonial legacy. Consequently, these writers have made the critique of homophobic and sexist ideologies an integral component of what we might call a decolonized Caribbean discourse.

CLAUDE MCKAY AND THE CONSTRUCTION OF (UN)NATURAL SEXUALITIES

A pioneering Jamaican writer who migrated to the United States in 1912, Claude McKay was, needless to say, a product of his time. In his biography of McKay, Wayne Cooper notes that although there is ample evidence to confirm his homosexuality, McKay never publicly identified himself as a homosexual and "like many homosexual writers of his day, did not seriously challenge the rule that such subjects were not to be discussed openly in creative literature."[10] Therefore, it is not surprising that McKay's most successful narratives—*Home to Harlem* (1928), *Banjo* (1929), and *Banana Bottom* (1933)—do not deal, at least in any explicit way, with the subject of homosexuality or contain any overt homosexual characters. On the contrary, *Home to Harlem* and *Banjo* both feature swaggering, good-natured, hyper-masculine protagonists, who are emphatically and unequivocally heterosexual. Passionate, sensual, and instinctive, Jake and Banjo embody the African American folk spirit that the narratives celebrate, representing what Bernard Bell calls "romantic prototypes of the rootlessness, creativity, and spiritual resilience of the common people of the race."[11]

Nevertheless, *Home to Harlem* and *Banjo* construct what is, in effect, a "homosocial" world of men interacting predominantly with other men.[12] This exclusively male domain is defined by the gamblers, musicians, hustlers, sailors, soldiers, Pullman porters, cooks, and waiters who typically populate McKay's novels. Although women are frequently objects of the protagonists' sexual desire—Jake's "tantalizing brown" in *Home to Harlem,* for example—the values and codes of this masculine domain are the ones that Jake and Banjo must strive to uphold and that the novels ultimately reinscribe and celebrate. Despite the vitality and passion with which McKay typically imbues his protagonists, the forms of masculinity that the narratives inscribe do not ultimately depart from traditional notions of maleness and masculine behavior. Indeed, McKay's folk heroes reflect and even reinforce dominant sexual ideologies by asserting a masculinity that is predicated on both sexism and homophobia. For example, during one of his stints as a cook working in a railroad dining car, Jake encounters a waiter reading a "French" (clearly a code for homosexual) novel. While questioning the waiter about the book—a story by Alphonse Daudet entitled *Sappho*—he begins to hum a tune that makes explicit the link between the novel's particular figuration of masculine identity and the sexist and homophobic values on which it depends: "And it is ashes to ashes and dust to dust, / Can you show me a woman a man can trust? / And there is two things in Harlem I don't understan' / It is a bulldycking woman and a faggotty man."[13]

McKay's participation in a discourse of primitivism that prevailed in both black and white literary circles of the era resulted in the replication of certain essentialist notions about blackness and black sexuality in particular. Reflecting tendencies that were more or less prevalent in the major literary and cultural movements that distinguished the period—the Harlem Renaissance in the United States and Négritude in Africa and the Caribbean, for example—McKay's texts constructed a notion of blackness that reinscribed a racial binary in which blacks were once again associated—albeit in a positive sense—with the realm of instincts, emotions, and passions, with sensuality, sexuality, and all that was considered "natural."

McKay's depictions of black urban life undoubtedly worked to disrupt class-bound notions of appropriateness and respectability—a fact that the negative response of certain black intellectuals to his work only serves to confirm.[14] In addition, his novels broke new ground in the sense that they challenged the taboo surrounding the representation of black sexuality. At the same time, however, his reinscription of a racial binary—especially one that depends so crucially on a category of the "natural"—implicitly articulates the very terms that have historically been used not only to devalue black cultures but also to marginalize gay and lesbian sexualities. The Caribbean feminist scholar M. Jacqui Alexander has demonstrated, for example, how the "naturalization" of heterosexuality as state law has traditionally depended on the designation of gay and lesbian sex as "unnatural." Furthermore, Alexander points out that "there is no absolute set of commonly understood or accepted principles called the 'natural' which can be invoked definitionally except as they relate to what is labelled 'unnatural.'"[15]

The reinscription of this category of the natural and its implicit corollary the unnatural is implicated even more explicitly in McKay's third novel, *Banana Bottom*, which is set in his native Caribbean. McKay's narrative is structured around a series of oppositions that include the native versus the European, Obeah versus Christianity, the primitive versus the civilized, instinct versus intellect, folic culture versus high culture, spontaneous warmth versus cultivated refinement, natural growth versus artificial growth, and so on. Within this schema, the female protagonist, Bita Plant—a name which is clearly meant to suggest the character's rootedness in the "native" soil of Jamaican folk culture—represents the triumph of "indigenous" cultural values over the metropolitan ones that have been imposed upon her. Consequently, Bita maintains an inherent and "instinctive" connection to the language, culture, and folkways of the rural peasantry from which she springs, despite her European education and the "seven years of polite upbringing" provided by the Craigs, the British missionary couple who adopted Bita and sent her to England.

Moreover, McKay's valorization of indigenous culture also entails the affirmation of a "native" sexuality, specifically coded as natural and therefore necessarily counterposed to the possibility of an unnatural or "aberrant" sexuality. Bita's marriage at the end of the novel to Jubban the drayman—instead of to Herald Newton Day, the Craigs's choice for Bita—signals the triumph of this natural sexuality as much as it represents the affirmation of an indigenous Jamaican folk culture. Like Jordan Plant, Bita's father—over whose literal dead body Bita and Jubban consummate their love—Jubban "possessed a deep feeling for the land" and he was "a lucky-born cultivator."[16] Jubban's sexual desires for Bita—and hers for him—are thus associated with the natural cycles of birth and death, growth and decay that determine the rhythms of peasant life. Furthermore, the sexuality that their union affirms is consequently linked to the reproductive laws that supposedly govern nature as well as humankind.

In contrast to Jubban's "natural" sexuality, Herald Newton Day, the promising young deacon who tragically "defiled himself with a nanny goat," represents the epitome of an "unnatural" sexuality. Whether explained as an "aberration" within nature as Teacher Fearon suggests or "the result of too much exclusive concentration on sacred textbooks and holy communion" as Squire Gensir conjectures (176-77), Herald's behavior constitutes a devia-tion from the (reproductive and heterosexual) norm that defines the "instinctive" sexuality of the black peasantry. In addition to reinforcing notions about racial atavism that circulate throughout the text, Herald's "aberrant" behavior also serves to confirm the novel's premise of the potentially degenerating effects of an overcivilized, sexually repressed, Western (European) civilization that privileges intellect over instinct, reason over emotions.

McKay's construction of a dichotomy between "natural" and "unnatural" sexualities consequently fixes "native" sexuality within certain narrow terms—restricting it to an exclusively reproductive function, for example—at the same time that it seems to link aberrant or unnatural sexual behavior (bestiality, rape, and presumably other forms of non-procreative sex) to the effects of either miscegenation or foreign "decadence and degeneracy."[17] Indeed, McKay's depiction of Squire Gensir—the eccentric Englishman who befriends Bita—prefigures, to a certain extent, the representation (in Paule Marshall's *The Chosen Place, The Timeless People* [1969], for example) of homosexuality and the homosexual as products of foreign "contamination."

As Cooper points out in his biography, Squire Gensir represents the "fictional prototype" for Walter Jekyll—the eccentric Englishman who served as one of McKay's literary patrons. Although Cooper acknowledges Jekyll's homosexuality, he cautiously asserts that in all probability "Jekyll's admiration and love... expressed itself wholly in his role as mentor and friend."[18] Given the unspoken taboo that in McKay's time precluded the explicit representation of homosexual characters, it is certainly not surprising that Jekyll's homosexuality is sublimated in the portrayal of Squire Gensir. Nevertheless, the traces of this repressed homosexuality are discernible in Gensir's so-called "eccentricity," his life-long bachelorhood, and his admission that he was "not a marrying man."[19] Indeed, the complete desexualization of Gensir within the novel underscores—by way of its conspicuous absence—what the narrative is unable to name. According to the text, Gensir "lived aloof from sexual contact" and was, as Mrs. Craig often remarked, "a happy old bachelor with... not the slightest blemish upon his character — a character about which nothing was whispered either *naturally* or otherwise" (emphasis added; 92). To the extent that Gensir remains an outsider whose appreciation of Jamaican folk culture is ultimately "merely cerebral" (85), the homosexuality (however latent) implied in his characterization is likewise encoded as non-native and therefore "foreign."

(NEO)COLONIALISM AND (HOMO)SEXUALITY IN PAULE MARSHALL'S THE CHOSEN PLACE, THE TIMELESS PEOPLE

In an exchange that in many ways echoes the cultural politics of the Buju Banton controversy, Hortense Spillers takes issue with Judith Fetterley's claim that Paule Marshall's novel, *The Chosen Place, The Timeless People,* is "homophobic." Fetterley's allegation is presumably based on her reading of the brief lesbian affair that takes place between Merle, the novel's protagonist, and a wealthy white woman who serves as her "London patroness." In an attempt to account for the divergence between her interpretation and that of Fetterley, Spillers suggests that the disagreement represents "an illustration of the sorts of conflicts that arise among discontinuous reading and interpretive communities." In addition, Spillers argues that Merle's lesbian encounter is not "a major thematic issue in the novel" and suggests instead that Marshall is more concerned in the episode with "the particular dynamics of colonial politics and its involvement on the intimate ground of feeling."[20]

Although Spillers's reading of the ways in which the relationship between Merle and her London patroness reflects the inequities of the colonial relation is certainly astute, the question of the novel's "homophobia" is not (or should not be) so easily dismissed. Rather than choosing between readings that emphasize either colonial *or* sexual politics, I would argue that the two are inextricably linked in Marshall's text and that Merle's encounter with the white lesbian functions as a trenchant critique of colonialism at the same time that it reinscribes certain dominant sexual ideologies. In fact, I would argue that this particular conjunction of the sexual and the colonial in Marshall's novel reflects the terms in which anti-colonial arguments were often constructed in certain "Afrocentric" or black nationalist discourses that characterized the period. Consequently, Marshall's formulation demonstrates how such discourses—especially insofar as they rely on notions of family or "race" as family—are always already gendered, always already, in Stuart Hall's words, "underpinned by a particular sexual economy, a particular figured masculinity [or femininity], a particular class identity," and so on.[21]

Although the "lesbian episode" may not appear to occupy a central place in the thematic scheme of the novel—Spillers points out, for example, that the encounter is only retrospectively recalled by Merle—it can (and should) be read alongside other episodes where the question of homosexuality is either implicitly or explicitly raised. A pattern of representation might, thus, be established in terms of the recirculation of certain ideologies of gender and sexuality within the narrative. These ideologies have to do not only with positioning gay and lesbian sexuality as foreign and/or unnatural but also with prescribing normative boundaries for male and female gender identity in general.

The affair between Merle and her London patroness is clearly meant to signify the asymmetries of the colonial relationship itself. This link between sexual and imperial motives is made explicit when Merle recalls her patroness's preference for "foreigners": "During the time I lived there I met people from every corner of the globe: India, Asia, Africa... all over the place. The sun, you might say, never set on the little empire she had going in her drawing room."[22] In addition, Merle's account of the indebtedness and dependency that the wealthy white woman would deliberately and strategically encourage exposes one of the primary mechanisms by which post-independence Caribbean states—like the fictitious Bourne Island— are kept under the crush of the neocolonial heel. However, at the same time that the portrayal of Merle's encounter with the English woman enacts an insightful critique of (neo) colonial politics, it simultaneously reinscribes a rhetoric that positions gay and lesbian sexuality as unnatural and foreign. Describing the "wild crowd" she fell in with, Merle states that they (the English) were "experts at making anything they do seem perfectly *natural*, and getting you to think so, too." In addition, she describes her patroness as "one of those upperclass types you hear of over there who don't seem to mind having produced a *degenerate* or two" (emphases added; 327-28).

This association of gay and lesbian sexuality with the unnatural also informs Marshall's characterization of the gay tourists who frequent Sugar's, the local nightclub. Like Merle's patroness, the gay men are portrayed as predators; their exploitation of "native" sexuality serves as an emblem of the economic exploitation that defines the neocolonial regime. Pointing out this group of affluent gay white men to Saul Amron, the white protagonist and later Merle's lover, Merle exclaims: "As for that bunch out on the balcony... Not a boy child over the age of three is safe since they arrived on the island" (87). Nevertheless, at the same time that it reflects an acute and subtle understanding of the way that the colonial dynamic permeates all levels of indigenous life—including what Spillers calls "the intimate ground of feeling"—this characterization also reinforces certain stereotypical notions about the unnaturalness of gay sexuality. The narrator states that these men "had the overstated gestures of their kind, as well as the *unnaturally* high voices that called attention to themselves and the laugh that was as shrill and sexless as a eunuch's, and which never ceased" (emphasis added; 88).

As M. Jacqui Alexander points out, these narrative figurations that position gay and lesbian sexuality as unnatural also serve to naturalize heterosexuality as an implicit norm.[23] In addition to ensnaring her in a cycle of debt and dependency, Merle's liaison with the English lesbian has the effect of destabilizing her identity as a woman. Merle admits that the "business between her and myself... had me so I didn't know who or what I was." And she confesses to Saul that when she finally decided to sever the ties with her patroness it was because "most of all... I was curious to see if a man would maybe look at me twice."[24] In other words, Merle's recuperation of a stable black female identity seems to hinge on her ability to attract the sexual attentions of a (heterosexual) male. In fact, Merle is eventually "saved" from the corrupting influence of the white lesbian not only by a man, but also through marriage and motherhood—in other words, the type of sexual relationship that epitomizes the heterosexual norm, what Alexander calls "conjugal heterosexuality"[25] Recalling her brief marriage to Ketu—the committed Ugandan nationalist she met in London—Merle states that "most of all, he made me know I was a woman.... After years of not being sure what I was, whether fish or fowl or what, I knew with him I was a woman and no one would ever again be able to make me believe otherwise. I still love him for that."[26] In the novel, Merle's lesbian affair represents a betrayal not only of her "true womanhood" —the "great wrong" that Ketu finds it impossible to forgive—but also of her anti-colonial politics, her family, and ultimately her "race." Afraid that Merle's touch might somehow "contaminate" their child, Ketu eventually abandons Merle and returns to Africa, taking their daughter with him. Demonstrating the degree to which she has internalized the supposed naturalness of the heterosexual norm, Merle herself is convinced that Ketu's actions are entirely justified. From another perspective, one might convincingly argue that Ketu leaves primarily because the knowledge of Merle's lesbian affair threatens his own sense of masculinity.

Similarly, Marshall's portrait of Allen Fuso, Saul's young assistant, implicitly assumes the universality or "naturalness" of a normative heterosexual masculinity. Although Allen's (latent) homosexuality is more complexly delineated than that of Merle's London patroness or the gay men at Sugar's, the resolution (or the lack thereof) of his "identity crisis" ultimately reveals the narrative's refusal to imagine anything other than a heterosexual solution to his "problem." Allen's crisis is precipitated by the homosexual feelings that are occasioned by his growing friendship with Vere, the "native son" who has returned home to Bourne Island after a brief stint on a labor scheme in the United States. Moreover, Allen's homosexuality is represented in the narrative as a kind of arrested development that is the consequence of an unresolved castration anxiety. Allen is unable to perform (hetero)sexually with Elvita—the date that Vere arranges for him during Carnival—not only because he finds women's bodies which "lacked purity of line with the up-jutting breasts and buttocks" distasteful, but also because of "his fear, borne of a recurrent phantasy of his as a boy, that once he entered that dark place hidden away at the base of their bodies, he would not be able to extricate himself" (309).

Alone on the settee, with Vere and Milly making love on the other side of the screen, Allen performs a solitary sexual act that signifies in the text not only a substitute for intercourse but also a parody of the (heterosexual) sex act itself. As Allen masturbates to the sounds of the unseen lovers, the image of Vere looms large in his imagination: "The girl was faceless, unimportant, but he saw Vere clearly: his dark body rising and falling, advancing and retreating, like one of the powerful Bournehills waves they sometimes rode together in the early evening" (312). Allen, in effect, erases Milly from the scene and puts himself in her place, thus becoming what he has subconsciously longed to be—namely, Vere's lover. "His cry at the end, which he tried to stifle but could not, broke at the same moment the girl uttered her final cry, and the two sounds rose together, blending one into the other, becoming a single complex note of the most profound pleasure and release" (312).

Given the ambiguity of the text—especially where issues of gay and lesbian sexuality are concerned—it is difficult to determine the extent to which Allen is aware (consciously, at least) of his own homosexuality. Nevertheless, he admits to Merle—in an attempt to explain the deep depression that had overtaken him since the events of Carnival and the subsequent death of Vere—that he longs for "something that wasn't so safe and sure all the time" or even "something people didn't approve of so they no longer thought of me as such a nice, respectable type." The obvious inadequacy of Merle's response —she recommends "a nice girl and some children"—is not surprising, given the guilt she has internalized as a result of her own lesbian affair (378-81). However, her response also reflects the novel's investment in and recirculation of an ideology that naturalizes heterosexuality while it positions gay and lesbian sexuality as deviant or unnatural. Furthermore, Merle's inability to imagine anything other than a conventional heterosexual (and reproductive) solution to Allen's "problem" not only defines the limits of the novel's discourse on questions of homosexuality, it also exposes one of the consequences—inherent in certain black nationalist discourses, for example—of uncritically conflating race with notions (especially naturalized ones) of family.

MICHELLE CLIFF, H. NIGEL THOMAS, AND THE CONTRADICTIONS OF REPRESENTING THE "INDIGENOUS" GAY/LESBIAN SUBJECT

> We are always in negotiation, not with a single set of oppositions that place us always in the same relation to others, but with a series of different positionalities. Each has for us its point of profound subjective identification.

> And that is the most difficult thing about this proliferation
> of the field of identities and antagonisms: they are. often
> dislocating in relation to one another.—STUART HALL,
> "What Is This 'Black' in Black Popular Culture?"

In an essay in which he attempts to map the critical challenges presented by the current historical conjuncture, Stuart Hall suggests that "it is to the diversity, not the homogeneity, of black experience that we must now give our undivided creative attention." Hall argues that, given the emergence of what he refers to as "a new kind of cultural politics," it is necessary now more than ever to "recognize the other kinds of difference [those of gender, sexuality, and class, for example] that place, position, and locate black people."[27] Two recent Caribbean writers—Michelle Cliff (Jamaica) and H. Nigel Thomas (St. Vincent)—have produced texts which reflect the way anti-colonial/imperial discourses need to be conceptualized in the context of the present historical and cultural situation. Attending to the differences that operate within as well as between cultures, these texts simultaneously critique the sexist/homophobic and colonial/neocolonial structures that continue to pervade contemporary Caribbean societies. By posing an implicit challenge to the binary oppositions that often define discussions of "native" sexuality, writers like Cliff and Thomas have cleared a discursive space for the articulation of an "indigenous" gay/lesbian subjectivity.

In contrast to these binary structures—which often imply the mutually exclusive choice of an either/or—these writers frequently deploy narrative strategies that privilege ambiguity and the ability to negotiate contradictions. For example, in an essay entitled "If I Could Write This in Fire, I Would Write This in Fire," Michelle Cliff relates an incident that underscores the contradictions generated by a Caribbean lesbian identity—contradictions that illustrate, in Hall's terms, how the multiple "positionalities" that inevitably constitute such an identity "are often dislocating in relation to one another." Cliff becomes justifiably enraged when she and Henry, a distant cousin who is "recognizably black and speaks with an accent," are refused service in a London bar. Although she is light-skinned enough to "pass" for white, Cliff states that she has "chosen sides." However, the lines suddenly become blurred—and allegiances begin to shift—when her cousin joins his white colleagues in a "sustained mockery" of the waiters in a gay-owned restaurant.[28] The conflicting feelings of anger (at Henry's homophobia/sexism) and solidarity (because he is also a victim of racism and colonial oppression) that exemplify Cliff's response to Henry mirror the profound ambivalence she feels towards Jamaica itself—the "killing ambivalence" that comes with the realization that home (especially for the "lesbian of color") is often a site of alienation as well as identification.[29]

This ambivalence is also reflected in the dual strategy that informs Cliff's first *novel*, Abeng (1984). On one hand, the narrative affirms the value of an indigenous Jamaican culture—especially the oral traditions and folk practices that embody the island's long history of anti-colonial resistance. On the other, the novel elaborates an incisive critique of the oppressive ideological structures that continue to pervade the postcolonial state—a deeply entrenched color-caste system, homophobia, and sexism, for example. Exemplifying the formal and stylistic innovations that are characteristic of her work, Cliff deliberately disrupts the narrative continuity of *Abeng* by intercutting the story of Clare Savage—the novel's young female protagonist—with fragments of history, myth, and legend. In her attempt to reconstruct what the critic Simon Gikandi calls a "repressed Afro-Caribbean history,"[30] Cliff inscribes a revisionary account that challenges not only the Eurocentric premises of conventional historiography but also its phallocentric and heterosexist assumptions as well. In other words, in addition to representing a female-centered tradition of resistance, *Abeng* also attempts to posit a historical or "genealogical" precedent for an indigenous lesbian/gay subjectivity.

Although it would perhaps be a historical misnomer to label Mma Alli—the mythical figure who plays a part in the novel's reconstruction of Caribbean slave resistance—a lesbian character per se, she clearly represents the possibility of an indigenous or even Afrocentric precedent for a non-heterosexual orientation:

> Mma Alli had never lain with a man. The other slaves said she loved only women in that way, but that she was a true sister to the men—the Black men: her brothers. They said that by being with her in bed, women learned all manner of the magic of passion. How to become wet again and again all through the night. How to soothe and excite at the same time. How to touch a woman in her deep-inside and make her womb move within her.[31]

Descended from a line of "one-breasted warrior women," Mma Alli is spiritually if not biologically related to Maroon Nanny—the slave leader who, according to local legend, "could catch a bullet between her buttocks and render the bullet harmless" (14)—and all the other female figures who function as historical precursors in a tradition from which Clare ("colonized child" that she is) has become tragically disconnected.

In addition to inscribing a proto-lesbian figure within the reconstructed mythology of an Afro-Caribbean past, Cliff exposes the homophobia that results in the marginalization and persecution of lesbians and gay men within contemporary Jamaican culture. These deeply ingrained homophobic attitudes—which reflect a fear of "difference"—represent one of the primary means by which a normative heterosexuality is consolidated and, indeed, enforced. For example, the story of Clinton, the son of "Mad Hannah," demonstrates what can happen if one is even suspected of being homosexual.

When Clinton is "taken with a cramp while... swimming in the river," he is left to drown while "shouts of 'battyman, battyman' echoed off the rocks and across the water of the swimming hole" (63). Likewise, the fate of an uncle who was rumored to be "funny" serves as an implicit warning to Clare against the dangers of transgressing the boundaries of what is culturally sanctioned as acceptable or "normal" sexual behavior. Although Clare was "not sure what 'funny' meant," she "knew that Robert had caused some disturbance when he brought a dark man home from Montego Bay and introduced him to his mother as 'my dearest friend.'" Stigmatized and ostracized by his family, Robert finally "did what Clare understood many 'funny' 'queer' 'off' people did: He swam too far out into Kingston Harbor and could not swim back. He drowned just as Clinton—about whom there had been similar whispers—had drowned" (125-26).

In Cliff's second novel, *No Telephone to Heaven* (1987), the ambivalence of the Caribbean gay/lesbian subject is literally embodied by the character Harry/ Harriet—the "boy-girl" who serves as the friend, confidant, and alter ego of an older Clare Savage. In the very indeterminacy of his/her name, Harry/Harriet reflects the unresolved (and perhaps unresolvable) contradictions that are inevitably generated by an indigenous gay or lesbian identity.

278 | CHAPTER TWELVE

Constantly transgressing the boundaries that supposedly separate male from female, upper from lower classes, insider from outsider, self from "other," natural from unnatural sexuality, Harry/Harriet inhabits an "interstitial" space—designated by the conjunction "both/and" rather than "either/or"—that, as he/she asserts, is "not just sun, but sun and moon."[32] In addition, Cliff clearly disrupts the discursive positioning of homosexuality as a "foreign contamination" by de-allegorizing the rape of Harry/Harriet when he was a child by a British officer. Although Harry/Harriet admits to Clare that he/she is often tempted to think "that what he [the officer] did to me is but a symbol for what they did to all of us " he/she asserts that the experience was not the "cause" of his ambiguous sexuality. Instead, Harry/Harriet insists on the concrete and literal brutality of the rape: "Not symbol, not allegory... merely a person who felt the overgrown cock of a big whiteman pierce the asshole of a lickle Black bwai" (129-30).

In his first novel *Spirits In The Dark* (1993), H. Nigel Thomas deploys a narrative construct that functions—especially in its Utopian gestures—much like the band of "revolutionaries" that Clare joins in *No Telephone to Heaven*. Jerome Quashee, the protagonist of Thomas's narrative, is initiated into an obscure and vaguely Afrocentric religious sect known as the Spiritualists. As a consequence, Jerome undergoes a ritual experience during the course of the novel that ultimately transforms and redeems him. Moreover, this redemptive experience becomes a way for Thomas to imagine and represent what might be called a decolonized Caribbean reality. In his attempt to articulate the ideological conditions of this decolonized Caribbean reality, Thomas insists on the need to dismantle not only oppressive political structures but restrictive sexual ones as well. Demonstrating an implicit understanding of how, as Stuart Hall puts it, "a transgressive politics in one domain is constantly sutured and stabilized by reactionary or unexamined politics in another,"[33] Thomas simultaneously confronts the patriarchal, heterosexist, and Eurocentric ideologies that constitute the particular legacy of the Caribbean colonial experience. Jerome's descent into "madness" —he suffers a series of "breakdowns" prior to his initiation— consequently reflects his unstable status as both a colonial subject and a homosexual.

Jerome's initiation entails a period of self-imposed isolation and sensory deprivation that enables him to reflect on and thereby come to terms with his experiences with a deeply flawed colonial school system, the expectations and disappointments of his parents, and his sexual feelings for other men. Assisted by Pointer Francis, who serves as his spiritual guide, Jerome emerges with a newfound understanding of his "African heritage" as well as an acceptance of his homosexuality. Jerome finally realizes that he had "put the sex part of [his] life 'pon a trash heap just fo' please society" and that "madness" was the price he paid for "hiding and sacrificing [Ms] life like that."[34]

However, if Jerome's spiritual rebirth constitutes a Utopian gesture that reflects Thomas's desire to inscribe a decolonized indigenous gay subject within his text, that gesture is necessarily tempered by the pervasive homophobia that the novel also exposes. Although Pointer Francis tells Jerome that there is "nothing sinful 'bout sex" even if Jerome is "a case of a pestle needing a pestle" (as opposed to a pestle needing a mortar or vice versa), he nevertheless reminds Jerome that he is "going back to live in the real world, with real people" and that "most o' the brethren ain't grown enough fo' understand why you is how yo' is and fo' accept yo' as yo' is" (198,212-13). In the course of his spiritual journey, Jerome recalls various incidents that were decisive in terms of his subconscious decision to repress his homosexuality. Chief among these are his memories of Boy Boy, the gay cousin who "was a constant point of reference for what the society would not accept" (94). The ridicule and humiliation that Boy Boy was forced to endure confirmed the unacceptability of Jerome's homosexual feelings. In addition, the fate that Boy Boy suffered when he "arranged with a young man to meet him in one of the canefields" demonstrated how physical violence was often used by the community in order to enforce a normative heterosexuality: "When he [Boy Boy] got there, there were ten of them. They took turns buggering him; one even used a beer bottle; then they beat him into unconsciousness and left him there. He'd refused to name the young men. But everyone knew who they were because they'd bragged about what they'd done—everything but the buggering" (199).

Jerome also recalls a more recent incident that illustrates how the community often acted in complicity with such violence by condoning or at least refusing to challenge these virulent displays of homophobic behavior. Jerome remembers that when Albert Brown, a cashier in the post office where he worked, was slapped by a coworker because he dared to offer a strikingly effective riposte to the mail sorter's crude homophobic insult, "no one, not even Jerome, reprimanded Brill." Moreover, when Jerome is called as a witness, the postmaster seems almost unwilling to believe his account, which leads Jerome to wonder if "perhaps the postmaster would have preferred that he he and save him from having to take action against Brill" (200). Despite the postmaster's apparent reluctance, Brill is eventually dismissed—it seems he was on probation at the time for "telling a female clerk that he didn't have to 'take orders from a cunt.'" Nevertheless, many of Jerome's coworkers were "angry with Albert, saying that he did not know how to take a joke as a man, that he had caused Brill to lose his job, and didn't he know that Brill had a wife and two children to feed?" Jerome is likewise criticized for not knowing "how to see and not see and hear and not hear" (200). The obvious implication is, of course, that homophobia and sexism are not considered serious offenses—since they uphold an apparently "natural" order —and therefore hardly warrant such severe censure. From this perspective, it is unimaginable that Brill would lose his job simply because "he slap a buller."

However, at the same time that it exposes the complicity of the community, Thomas's text, like Cliff's, demonstrates an acute sensitivity to the ambiguous and sometimes contradictory spaces that inevitably exist in any culture. These sites of ambiguity and contradiction—which often reflect how "differences" are actually lived and negotiated—are, paradoxically perhaps, the ones that can potentially enable new forms of social and cultural relations. For example, at one point in his meditations, Jerome finds himself contemplating an episode that reveals the surprising capacity for tolerance that also exists alongside the homophobia pervading all levels of Caribbean society. Jerome recalls that among the female food vendors who plied their trade in the open-air market, there was also "a man whom the buyers and non-buyers said was the biggest woman of the lot. They called him Sprat." Because he "got more customers than the women," Sprat often became the target of homophobic insults in the quarrels that frequently broke out as a result of the fierce competition among the vendors. However, despite the caustic nature of these exchanges, Jerome observes that Sprat nevertheless "loaned Melia [a vendor with whom he had previously argued] ten dollars to buy some ground provisions somebody was selling at a bargain." Noting Sprat's absence on another occasion, Jerome learns that he had the flu and that "three of the women had been to see him. One said he would be out the next week and she was buying supplies for him that day" (21-22). What Jerome comes to understand, then, is that there are relations of professional and personal reciprocity binding Sprat and the other vendors together—existing social relations which pose a contradiction to the homophobic ideologies that serve to position him as other.

Indeed, Thomas's novel seems to suggest that the willingness to accept the indeterminacy associated with such contradictions —the opposite of rigid binary thinking, in other words—is often the first step in undoing the homophobia that continues to marginalize lesbians and gay men in contemporary Caribbean cultures. For example, Pointer Francis reminds Jerome that it "is only when most people have a son or a daughter that is like that [that] they stop ridiculing and start thinking" (213). Once again, it is particularly within the context of concrete affiliative social relations that the potential for negotiating these contradictions can exist. Consequently, Jerome singles out his brother, Wesi, as the one "he would tell... everything about himself," mainly because Wesi "was the first person he knew that understood and accepted contradictions" (156). In fact, one of the central insights that Jerome gleans from his initiation has to do precisely with the importance of this "non-binary" mode of thinking: "Jerome knew that by the time the spirit called you, you knew that life itself was a contradiction" (177).

In the context of an indigenous criticism, the need for non-binary modes of thinking that resist the totalizing impulses implicit in both the "universalist" and "nativist" positions—the impasse between which the Buju Banton controversy so clearly exemplifies—is equally urgent. Given the alarming persistence of anti-gay violence in contemporary Caribbean societies and the reproduction in literature and popular culture of ideologies that condone or legitimate such violence, we clearly need a critical practice that goes beyond simple dichotomies—us/them, native/foreign, natural/unnatural—a practice that can not only affirm but also critique indigenous cultures in all of their varied and inevitably contradictory forms.

Notes

1   "Boom Bye Bye," written by Mark Myrie, *Buju Banton: The Early Years 90-95* (Kingston, Jamaica: Penthouse Record Distributors, 1992). See Ransdell Pierson, "'Kill Gays' Hit Song Stirs Fury" *New York Post*, October 24, 1992; Joan Morgan, "No Apologies, No Regrets," *VIBE*, October 1993, 76-82; Peter Noel and Robert Marriott, "Batty Boys in Babylon: Can Gay West Indians Survive the 'Boom Bye Bye' Posses?" *Village Voice*, January 12,1993.

2   Noel and Marriott, "Batty Boys in Babylon," 35, 31.

3   Morgan, "No Apologies, No Regrets," 76.

4   Ibid., 82.

5   Carolyn Cooper," 'Lyrical Gun': Metaphor and Role Play in Jamaican Dancehall Culture," *Massachusetts Review* 35, no. 3-4 (1994): 437. Cooper borrows the notion of the "lyrical gun" from Shabba Ranks's dancehall tune, "Gun Pon Me."

6   Ibid., 438.

7   "Open Letter," *VIBE*, October 1993, 82. This letter was published alongside Morgan's *VIBE* piece and collectively signed by many prominent "lesbians, gay men, and transgendered persons of African, Afro-American, Afro-Caribbean, and Afro-Latin descent."

8   The quote is taken from an unpublished paper that the author was kind enough to share with me: "On Liminal Subjectivity" (paper presented at the National Symposium on Indigenous Knowledge and Contemporary Social Issues, Tampa, Florida, March 3-5,1994).

9   According to Cooper, plans for a protest to be led by a group of "local homosexuals" failed to materialize because "on the day of the rumoured march, men of all social classes gathered in the square, armed with a range of implements—sticks, stones, machetes— apparently to defend their heterosexual honor." Nevertheless, Cooper states that the aborted attempt paradoxically generated a public discourse on homosexuality when, in the wake of the non-event, "numerous callers on various talk show programs aired their opinions in defence of, or attack on the homosexual's right to freedom of expression" ("Lyrical Gun," 440). In addition, Isaac Julien's film, *The Darker Side of Black* (Arts Council of Great Britain, 1993), explores these very issues in relation to rap, hip-hop, and African American popular culture in general, as well as its diasporic counterpart in the Caribbean, the culture of the dancehall.

10   Wayne Cooper, *Claude McKay: Rebel Sojourner in the Harlem Renaissance* (Baton Rouge: Louisiana State University Press, 1987), 75.

11   Bernard Bell, *The Afro-American Novel and Its Tradition* (Amherst: University of Massachusetts Press, 1987), 118.

12   I am indebted to my colleague, Charles Nero, for helping me to clarify this concept of the homosocial in McKay's novels.

13   Claude McKay, *Home to Harlem* (New York: Harper and Brothers, 1928), 129.

14   For example, see Wayne Cooper's discussion of the critical reception — especially on the part of black American intellectuals like W. E. B. Du Bois — of McKay's *Home to Harlem* (*Claude McKay*, 238-48).

15   M. Jacqui Alexander, "Not Just (Any) *Body* Can Be A Citizen: The Politics of Law, Sexuality and Postcoloniality in Trinidad and Tobago and the Bahamas," *Feminist Review*, no. 48 (1994): 9.

16   Claude McKay, *Banana Bottom* (1933; New York: Harcourt, Brace, Jovanovich, 1961),291.

17   Early in the novel, Bita is raped by Crazy Bow Adair, a third generation descendant of a "strange Scotchman who had emigrated to Jamaica in the eighteen-twenties." Although the narrator states that the mixed-race progeny of this "strange liberator" were, for the most part, "hardy peasants," he nevertheless admits that there are those who believe "the mixing of different human strains" had less salutary effects (2-4). In addition, the novel suggests that Patou, the "cripple-idiot" son of the missionary couple, Priscilla and Malcolm Craig, is both a product and a reflection of the repressed sexuality that is associated with their Englishness.

18   Cooper, *Claude McKay*, 32.

19   McKay, *Banana Bottom*, 126.

20   In a footnote to her essay, Spillers states that Fetterley discussed parts of the novel with her, and in Fetterley's opinion, "the work is homophobic." Spillers cites Fetterley's book, *The Resisting Reader: A Feminist Approach to American Fiction* (Bloomington: Indiana University Press, 1978), as an example of work done on the "evaluative dynamics of critical reading and the formation of reader communities." Hortense Spillers, "Chosen Place, Timeless People: Some Figurations on the New World," in *Conjuring: Black Women, Fiction, and Literary Tradition,* ed. Marjorie Pryse and Hortense Spillers (Bloomington: Indiana University Press, 1985), 172-74 n. 6.

21   Stuart Hall, "What Is This 'Black' in Black Popular Culture?" in *Black Popular Culture,* ed. Gina Dent (Seattle: Bay Press, 1992), 31. See also Paul Gilroy's article, "It's a Family Affair," in the same anthology, 303-16; and Anne McClintock's essay, "Family Feuds: Gender, Nationalism and the Family," *Feminist Review*, no. 44 (1993): 61-80.

22   Paule Marshall, *The Chosen Place, The Timeless People* (1969; New York: Vintage, 1992), 328.

23   Alexander, "Not Just (Any) *Body* Can Be A Citizen," 5-6.

24   Marshall, *The Chosen Place*, 329.

25   Alexander, "Not Just *(Any)Body* Can Be A Citizen," 10.

26   Marshall, *The Chosen Place*, 332.

27   Hall, "What Is This 'Black' in Black Popular Culture?" 30.

28   See p. 68 in Michelle Cliff, *The Land of Look Behind* (Ithaca, N.Y: Firebrand Books, 1985), 57-76.

29   Ibid., 103.

30   Simon Gikandi, *Writing in Limbo: Modernism and Caribbean Literature* (Ithaca, NY.: Cornell University Press, 1992), 233.

31   Michelle Cliff, *Abeng* (1984; New York: Penguin Books, 1991), 35.

32   Michelle Cliff, *No Telephone to Heaven* (1987; New York: Vintage, 1989), 171.

33   Hall, "What Is This 'Black' in Black Popular Culture ?" 31.

34   H. Nigel Thomas, *Spirits In The Dark* (1993; Oxford: Heinemann Publishers, 1994), 198.

CHAPTER THIRTEEN

Man Royals and Sodomites: Some Thoughts on the Invisibility of Afro-Caribbean Lesbians (1992)

MAKEDA SILVERA

I will begin with some personal images and voices about woman-loving. These have provided a ground for my search for cultural reflections of my identity as a Black woman artist within the Afro-Caribbean community of Toronto. Although I focus here on my own experience (specifially, Jamaican), I am aware of similarities with the experience of other Third World women of color whose history and culture has been subjected to colonization and imperialism.

I spent the first thirteen years of my life in Jamaica among strong women. My great-grandmother, my grandmother, and grand-aunts were major influences in my life. There are also men whom I remember with fondness—my grandmother's "man friend" G., my Uncle Bertie, his friend Paul, Mr. Minott, Uncle B., and Uncle Freddy. And there were men like Mr. Eden who terrified me because of stories about his "walking" fingers and his liking for girls under age fourteen.

I lived in a four-bedroom house with my grandmother, Uncle Bertie, and two female tenants. On the same piece of land, my grandmother had other tenants, mostly women and lots and lots of children. The big veranda of our house played a vital role in the social life of this community. It was on that veranda that I received my first education on "Black women's strength"—not only from their strength but also from the daily humiliations they bore at work and in relationships. European experience coined the term *feminism,* but the term *Black women's strength* reaches beyond Eurocentric definitions to describe the cultural continuity of my own struggles.

The veranda. My grandmother sat on the veranda in the evenings after all the chores were done to read the newspaper. People —mostly women-gathered there to discuss "life." Life covered every conceivable topic—economic, local, political, social, and sexual: the high price of salt-fish, the scarcity of flour, the nice piece of yellow yam bought at Coronation Market, Mr. Lam (the shopkeeper who was taking "liberty" with Miss Inez), the fights women had with their menfolk, work, suspicions of Miss Iris and Punsie carrying on something between them, the cost of school books.

My grandmother usually had lots of advice to pass on to the women on the veranda, all grounded in the Bible. Granny believed in Jesus, in good and evil, and in repentance. She was also a practical and sociable woman. Her faith didn't interfere with her perception of what it meant to be a poor Black woman; neither did it interfere with our Friday night visits to my Aunt Marie's bar. I remember sitting outside on the piazza with my grandmother, two grand-aunts, and three or four of their women friends. I liked their flashy smiles and I was fascinated by their independence, ease, and their laughter. I loved their names—Cherry Rose, Blossom, Jonesie, Poinsietta, Ivory, Pearl, Iris, Bloom, Dahlia, Babes. Whenever the conversation came around to some "big 'oman talk"—who was sleeping with whom or whose daughter just got "fallen"—I was sent off to get a glass of water for an adult, or a bottle of Kola Champagne. Every Friday night I drank as much as half a dozen bottles of Kola Champagne, but I still managed to hear snippets of words, tail ends of conversations about women together.

In Jamaica, the words used to describe many of these women would be *man royal* and/or *sodomite.* Dread words. So dread that women dare not use these words to name themselves. They were names given to women by men to describe aspects of our lives that men neither understood nor approved.

I heard "sodomite" whispered a lot during my primary-school years; and tales of women secretly having sex, joining at the genitals, and being taken to the hospital to be "cut" apart were told in the schoolyard. Invariably, one of the women would die. Every five to ten years the same story would surface. At times, it would even be published in the newspapers. Such stories always generated much talking and speculation from "Bwoy dem kinda gal naasti sah!" to some wise old woman saying, "But dis caan happen, after two shutpan caan join"—meaning identical objects cannot go into the other. The act of loving someone of the same sex was sinful, abnormal—something to hide. Even today, it isn't unusual or uncommon to be asked, "So how do two 'omen do it?... What unoo use for a penis?... Who is the man and who is the 'oman?" It's inconceivable that women can have intimate relationships that are whole, that are not lacking because of the absence of a man. It's assumed that women in such relationships must be imitating men.

The word *sodomite* derives from the Old Testament. Its common use to describe lesbians (or any strong, independent woman) is peculiar to Jamaica—a culture historically and strongly grounded in the Bible. Although Christian values have dominated the world, their effect in slave colonies is particular. Our foreparents gained access to literacy through the Bible when they were indoctrinated by missionaries. It provided powerful and ancient stories of strength, endurance, and hope which reflected their own fight against oppression. This book has been so powerful that it continues to bind our lives with its racism and misogyny. Thus, the importance the Bible plays in Afro-Caribbean culture must be recognized in order to understand the historical and political context for the invisibility of lesbians. The wrath of God "rained down burning sulphur on Sodom and Gomorrah" (Genesis 19:24). How could a Caribbean woman claim the name?

When, thousands of miles away and fifteen years after my school days, my grandmother was confronted with my love for a woman, her reaction was determined by her Christian faith and by this dread word *sodomite*—its meaning, its implication, its history.

And when, Bible in hand, my grandmother responded to my love by sitting me down, at the age of twenty-seven, to quote Genesis, it was within the context of this tradition, this politic. When she pointed out that "this was a white people ting," or "a ting only people with mixed blood was involved in" (to explain or include my love with a woman of mixed blood), it was a strong denial of many ordinary Black working class women she knew.

It was finally through my conversations with my grandmother, my mother, and my mother's friend five years later that I began to realize the scope of this denial which was intended to dissuade and protect me. She knew too well that any woman who took a woman lover was attempting to walk on fire—entering a "no man's land." I began to see how commonplace the act of loving women really was, particularly in working class communities. I realized, too, just how heavily shame and silence weighed down this act.

A conversation with a friend of my mother:

Well, when I was growing up we didn't hear much 'bout woman and woman. They weren't "suspect" There was much more talk about "battyman businesses" when I was a teenager in the 1950s.

I remember one story about a man who was "suspect" and that every night when he was coming home, a group of guys use to lay wait for him and stone him so viciously that he had to run for his life. Bern time, he was safe only in the day.

Now with women, nobody really suspected. I grew up in the country and I grew up seeing women holding hands, hugging-up, sleeping together in one bed, and there was no question. Some of this was based purely on emotional friendship, but I also knew of cases where the women were dealing, but no one really suspected. Close people around knew, but not everyone. It wasn't a thing that you would go out and broadcast. It would be something just between the two people.

Also one important thing is that the women who were involved carried on with life just the same; no big political statements were made. These women still went to church, still got baptized, still went on pilgrimage, and I am thinking about one particular woman name Aunt Vie, a very strong woman, strong-willed and everything; they use to call her "man royal" behind her back, but no one ever dare to meddle with her.

Things are different now in Jamaica. Now all you have to do is not respond to a man's call to you and dem call you sodomite or lesbian. I guess it was different back then forty years ago, because it was harder for anybody to really conceive of two women sleeping together and being sexual. But I do remember when you were "suspect" people would talk about you. You were

*definitely classed as "different" "not normal," a bit of a "crazy."
But women never really got stoned like the men.*

*What I remember is that if you were a single woman alone or
two single women living together and a few people suspected
this... and when I say a few people I mean like a few guys, sometimes other crimes were committed against the women. Some
very violent; some very subtle. Battery was common, especially
in Kingston. A group of men would suspect a woman or have
it out for her because she was a "sodomite" or because she act
"man royal" and so the men would organize and gang rape
whichever woman was "suspect." Sometimes it was reported
in the newspapers; other times it wasn't—but when you live in
a little community, you don't need a newspaper to tell what's
going on. You know by word of mouth and those stories were
frequent. Sometimes you also knew the men who did the battery.*

*Other subtle forms of this was "scorning" the women. Mean-
ing that you didn't eat anything from them, especially a cooked
meal. It was almost as if those accused of being "man royal" or
"sodomite" could contaminate.*

A conversation with my grandmother:

*I am only telling you this so that you can understand that this
is not a profession to be proud of and to get involved in. Everybody should be curious and I know you born with that, ever
since you growing up as a child, and I can't fight against that,
because that is how everybody get to know what's in the world.
I am only telling you this because when you were a teenager,
you always say you, want to experience everything and make
up your mind on your own. You didn't like people telling you
what was wrong and right. That always use to scare me.*

*Experience is good, yes. But it have to be balanced; you have
to know when you have too much experience in one area. I am
telling you this because I think you have enough experience in
this to decide now to go back to the normal way. You have two
children. Do you want them to grow up knowing this is the life
you-taken? But this is for you to decide...*

*Yes, there was a lot of women involved with women in Jamaica.
I knew a lot of them when I was growing up in the country in*

the 1920s. I didn't really associate with them. Mind you, I was not rude to them. My mother wouldn't stand for any rudeness from any of her children to adults.

I remember a woman we use to call Miss Bibi. She lived next to us—her husband was a fisherman, I think he drowned before I was born. She had a little wooden house that back onto the sea, the same as our house. She was quiet, always, reading. That I remember about her because she use to go to the little public library at least four days out of the week. And she could talk. Anything you wanted to know, just ask Miss Bibi and she could tell you. She was mulatto woman, but poor. Anytime I had any schoolwork that I didn't understand, I use to ask her. The one thing I remember though, we wasn't allowed in her house by my mother, so I use to talk to her outside, but she didn't seem to mind that. Some people use to think she was mad because she spent so much time alone. But I didn't think that because anything she help me with, I got a good mark on it in school.

She was colorful in her own way, but quiet, always alone, except when her friend come and visit her once a year for two weeks. Them times I didn't see Miss Bibi much, because my mother told me I couldn't go and visit her. Sometimes I would see her in the market exchanging and bartering fresh fish for vegetables and fruits. I use to see her friend, too. She was a jet Black woman, always her hair tied in bright colored cloth, and she always had on big gold earrings. People use to say she lived on the other side of the Island with her husband and children and she came to Port Maria once a year to visit Miss Bibi.

My mother and father were great storytellers and I learnt that from them, but is from Miss Bibi that I think I learnt to love reading so much as a child. It wasn't until I move to Kingston that I notice other women like Miss Bibi...

Let me tell you about Jones. Do you remember her? Well she was the woman who lived the next yard over from us. She is the one who really turn me against people like that and why I fear so much for you to be involved in this thing. She was very loud. Very show-off. Always dressed in pants and man-shift that she borrowed from her husband. Sometimes she use to invite me over to her house, but I didn't go. She always had her hair in a bob haircut, always barefoot and tending to her garden and her fruit trees. She tried to get me involved in that kind of life, but I said no. At the time I remember I needed some money to borrow and she lent me, later she told me I didn't have to pay her back, but to come over to her house and see the thing she had that was sweeter than what any man could offer me. I told her no and eventually paid her back the money.

We still continued to talk. It was hard not to like Jonesie—that's what everybody called her. She was open and easy to talk to. But still there was a fear in me about her. To me it seem like she was in a dead end with nowhere to go. I don't want that for you.

I left my grandmother's house that day feeling anger and sadness for Miss Jones—maybe for myself, who knows? I was feeling boxed in. I had said nothing. I'd only listened quietly.

In bed that night, I thought about Miss Jones. I cried for her (for me) silently. I remembered her, a mannish-looking Indian woman with flashy gold teeth, a Craven A cigarette always between them. She was always nice to me as a child. She had the sweetest, juiciest Julie, Bombay, and East Indian mangoes on the street. She always gave me mangoes over the fence. I remember the dogs in her yard and the sign on her gate. "Beware of bad dogs." I never went into her house, although I was always curious.

I vaguely remember her pants and shirts, although I never thought anything of them until my grandmother pointed them out. Neither did I recall that dreaded word being used to describe her, although everyone on the street knew about her.

A conversation with my mother:

Yes, I remember Miss Jones. She smoke a lot, drank a lot. In fact, she was an alcoholic. When I was in my teens she use to come over to our house—always on the veranda. I can't remember her sitting down—seems she was always standing up, smoking, drinking, and reminiscing. She constantly talked about the past, about her life. And it was always women: young women she knew when she was a young woman, the fun they had together and how good she could make love to a woman. She would say to whoever was listening on the veranda, "Dem girls I use to have sex with was shapely. You shoulda know me when I was younger, pretty, and shapely just like the 'oman dem I use to have as my 'oman."

People use to tease her on the street, but not about being a lesbian or calling her sodomite. People use to tease her when she was drunk, because she would leave the rumshop and stagger down the avenue to her house.

I remember the women she use to carry home, usually in the daytime. A lot of women from downtown, higglers and fish-women. She use to boast about knowing all kinds of women from Coronation Market and her familiarity with them. She had a husband who lived with her and that served her as her greatest protection against other men taking steps with her. Not that anybody could easily take advantage of Miss Jones; she could stand up for herself. But having a husband did help. He was a very quiet, insular man. He didn't talk to anyone on the street. He had no friends so it wasn't easy for anyone to come up to him and gossip about his wife.

No one could go to her house without being invited, but I wouldn't say she was a private person. She was a loner. She went to the rumshops alone, she drank alone, she staggered home alone. The only time I ever saw her with somebody were the times when she went off to the Coronation Market or some other place downtown to find a woman and bring her home. The only times I remember her engaging in conversation with anybody was when she came over on the veranda to talk about her women and what they did in bed. That was all she let out about herself. There was nothing about how she was feeling, whether she was sad or depressed, lonely, happy. Nothing. She seemed to cover up all of that with her loudness and her vulgar-ness and her constant threat—which was all it was—to beat up anybody who troubled her or teased her when she was coming home from the rumshop.

Now Cherry Rose—do you remember her? She was a good friend of Aunt Marie and of Mama's. She was also a sodomite. She was loud too, but different from Miss Jones. She was much more outgoing. She was a barmaid and had lots of friends—both men and women. She also had the kind of personality that attracted people—very vivacious, always laughing, talking, and touching. She didn't have any children, but Gem did.

Do you remember Miss Gem? Well, she had children and she was also a barmaid. She also had lots of friends. She also bad a man friend name Mickey, but that didn't matter because some women had their men and still had women they carried on with. The men usually didn't know what was going on, and seeing as these men just come and go and usually on their own time, they weren't around every day and night.

Miss Pearl was another one that was in that kind of thing. She was a dressmaker; she use to sew really good. Where Gem was light complexion, she was a very black Black woman with deep dimples. Where Gem was a bit plump, Pearl was slim, but with big breasts and a big bottom. They were both pretty women.

I don't remember hearing that word sodomite *a lot about them. It was whispered sometimes behind their back but never in front of them. And they were so alive and talkative that people were always around them.*

The one woman I almost forgot was Miss Opal, a very quiet woman. She use to befriends with Miss Olive and was always out of her bar sitting down. I can't remember much about her except she didn't drink like Miss Jones and she wasn't vulgar. She was soft-spoken, a half-Chinese woman. Her mother was born in Hong Kong and her father was a Black man. She could really bake. She use to supply shops with cakes and other pastries.

So there were many of those kind of women around. But it wasn't broadcast.

I remembered them. Not as lesbians or sodomites or man royals, but as women that I liked. Women whom I admired. Strong women, some colorful, some quiet.

I loved Cherry Rose's style. I loved her loudness, the way she challenged men in arguments, the bold way she laughed in their faces, the jingle of her gold bracelets. Her colorful and stylish way of dressing. She was full of wit; words came alive in her mouth.

Miss Gem: I remember her big double iron bed. That was where Paula and Lorraine (her daughters, my own age) and I spent a whole week together when we had chicken pox. My grandmother took me there to stay for the company. It was fun. Miss Gem lived right above her bar, and so at any time we could look through the window and on to the piazza and street, which was bursting with energy and life. She was a very warm woman, patient and caring. Every day she would make soup for us and tell us stories. Later on in the evening she would bring us Kola Champagne.

Miss Pearl sewed dresses for me. She hardly ever used her tape measure—she could just take one look at you and make you a dress fit for a queen. What is she doing now, I asked myself? And Miss Opal, with her calm and quiet, where is she—still baking?

What stories could these lesbians have told us? I, an Afro-Caribbean woman living in Canada, come with this baggage—their silenced stories. My grandmother and mother know the truth, but silence still surrounds us. The truth remains a secret to the rest of the family and friends, and I must decide whether to continue to sew this cloth of denial or break free, creating and becoming the artist that I am, bringing alive the voices and images of Cherry Rose, Miss Gem, Miss Jones, Opal, Pearl, and others....

There is more at risk for us than for white women. Through three hundred years of history we have carried memories and the scars of racism and violence with us. We are the sisters, daughters, mothers of a people enslaved by colonialists and imperialists.

Under slavery, production and reproduction were inextricably linked. Reproduction served not only to increase the labor force of slave owners but also, by "domesticating" the enslaved, to facilitate the process of social control. Simultaneously, the enslaved responded to dehumanizing conditions by focusing on those aspects of life in which they could express their own desires. Sex was an area in which to articulate one's humanity, but because it was tied to attempts "to define oneself as human," gender roles, as well as the act of sex, became badges of status. To be male was to be the stud, the procreator; to be female was to be fecund, and one's femininity was measured by the ability to attract and hold a man and to bear children. In this way, slavery and the post-emancipated colonial order defined the structures of patriarchy and hetero-sexuality as necessary for social mobility and acceptance.

Socioeconomic conditions and the quest for a better life have driven steady migration from Jamaica and the rest of the Caribbean to the United States, Britain, and Canada. Upon my arrival, I became part of the "visible minorities," encompassing Blacks, Asians, and Native North Americans in Canada. I live with a legacy of continued racism and prejudice. We confront this daily, both as individuals and as organized political groups. Yet for those of us who are lesbians, there is another struggle: the struggle for acceptance and positive self-definition within our own communities. Too often, we have had to sacrifice our love for women in political meetings that have been dominated by the "we are the world" attitude of heterosexual ideology. We have had to hide too often that part of our identity which contributes profoundly to the whole.

Many lesbians have worked, like me, in the struggles of Black people since the 1960s. We have been on marches every time one of us gets murdered by the police. We have been at sit-ins and vigils. We have flyered, postered; we have cooked and baked for the struggle. We have tended to the youths. And we have all at one time or another given support to men in our community, all the time painfully holding on to, obscuring, our secret lives. When we do walk out of the closet (or are thrown out), the "ideologues" of the Black communities say "Yes, she was a radical sistren, but I don't know what happened; she just went the wrong way." What is implied in this is that one cannot be a lesbian and continue to do political work and, not surprisingly, it follows that a Black lesbian/ artist cannot create using the art forms of our culture. For example, when a heterosexual male friend carne to my house, I put on a dub poetry tape. He asked, "Are you sure that sistren is a lesbian?"

"Why?" I ask.

"Because this poem sound wicked; it have lots of rhythm; it sounds cultural."

Another time, another man commented on my work, "That book you wrote on domestic workers is really a fine piece of work. I didn't know you were that informed about the economic politics of the Caribbean and Canada." What are we to assume from these statements? That Afro-Caribbean lesbians have no Caribbean culture? That they lose their community politics when they sleep with women? Or that Afro-Caribbean culture is a heterosexual commodity?

The presence of an "out" Afro-Caribbean lesbian in our community is dealt with by suspicion and fear from both men and our heterosexual Black sisters. It brings into question the assumption of heterosexuality as the only "normal" way. It forces them to acknowledge something that has always been covered up.

It forces them to look at women differently and brings into question the traditional Black female role. Negative responses from our heterosexual Black sisters, although more painful, are, to a certain extent, understandable because we have no race privilege and very, very few of us have class privilege. The one privilege within our group is heterosexual. We have all suffered at the hands of this racist system at one time or another, and to many heterosexual Black women it is inconceivable, almost frightening, that one could turn her back on credibility in our community and the society at large by being lesbian. These women are also afraid that they will be labeled "lesbian" by association. It is that fear, that homophobia, which keeps Black women isolated.

The Toronto Black community has not dealt with sexism. It has not been pushed to do so. Neither has it given a thought to its heterosexism. In 1988, my grandmother's fear is very real, very alive. One takes a chance when one writes about being an Afro-Caribbean lesbian. There is the fear that one might not live to write more. There is the danger of being physically "disciplined" for speaking as a woman-identified woman.

And what of our white lesbian sisters and their community? They have learned well from the civil rights movement about organizing, and with race and some class privilege, they have built a predominantly white lesbian (and gay) movement—a precondition for a significant body of work by a writer or artist. They have demanded and received recognition from politicians (no matter how little). But this recognition has not been extended to Third World lesbians of color—neither from politicians nor from white lesbian (and gay) organizations. The white lesbian organizations and groups have barely (some not at all) begun to deal with or acknowledge their own racism, prejudice, and biases—all learned from a system which feeds on their ignorance and grows stronger from its institutionalized racism. Too often white women focus only on their oppression as lesbians, ignoring the more complex oppression of non-white women who are also lesbians. We remain outsiders in these groups, without images or political voices that echo our own. We know too clearly that, as non-white lesbians in this country, we are politically and socially at the very bottom of the heap. Denial of such differences robs us of true visibility. We must identify and define these differences and challenge the movements and groups that are not accessible to non-whites—challenge groups that are not accountable.

But where does this leave us as Afro-Caribbean lesbians, as part of this "visible minority" community? As Afro-Caribbean women we are still at the stage where we have to imagine and discover our existence, past and present. As lesbians, we are even more marginalized, less visible. The absence of a national Black lesbian and gay movement through which to begin to name ourselves is disheartening. We have no political organization to support us, through which we could demand respect from our communities. We need such an organization to represent our interests, both in coalition building with other lesbian and gay organizations, and in the struggles which shape our future—through which we hope to transform the social, political, and economic systems of oppression as they affect all peoples.

Although not yet on a large scale, lesbians and gays of Caribbean descent are beginning to seek each other out—are slowly organizing. Younger lesbians and gays of color are beginning to challenge and force their parents and the Black community to deal with their sexuality. They have formed groups, Zami for Black and Caribbean Gays and Lesbians and Lesbians of Color, to name two.

The need to make connections with other Caribbean and Third World people of color who are lesbian and gay is urgent. This is where we can begin to build that other half of our community, to create wholeness through our art. This is where we will find the support and strength to struggle, to share our histories and to record these histories in books, documentaries, film, sound, and art. We will create a rhythm that is uniquely ours—proud, powerful, and gay. Being invisible no longer. Naming ourselves, and talcing our space within the larger history of Afro-Caribbean peoples. A dream to be realized, a dream to act upon.